THE LIBRARY
ST. MARY'S COLLEGE OF MARYLAND
ST. MARY'S CITY, MARYLAND 20686

Chiral Separations by
CHROMATOGRAPHY

Chiral Separations by CHROMATOGRAPHY

SATINDER AHUJA

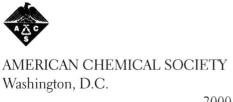

AMERICAN CHEMICAL SOCIETY
Washington, D.C.

2000

OXFORD
UNIVERSITY PRESS

Oxford New York
Athens Auckland Bangkok Bogotá Buenos Aires Calcutta
Cape Town Chennai Dar es Salaam Delhi Florence Hong Kong Istanbul
Karachi Kuala Lumpur Madrid Melbourne Mexico City Mumbai
Nairobi Paris São Paulo Singapore Taipei Tokyo Toronto Warsaw

and associated companies in
Berlin Ibadan

Copyright © 2000 by American Chemical Society

Developed and distributed in partnership by
American Chemical Society and Oxford University Press

Published by Oxford University Press, Inc.
198 Madison Avenue, New York, New York 10016

Oxford is a registered trademark of Oxford University Press

All rights reserved. No part of this publication may be reproduced,
stored in a retrieval system, or transmitted, in any form or by any means,
electronic, mechanical, photocopying, recording, or otherwise,
without the prior permission of American Chemical Society.

Library of Congress Cataloging-in-Publication Data
Ahuja, Satinder, 1933–
Chiral separations by chromatography / Satinder Ahuja.
p. cm.
ISBN 0-8412-3631-3
1. Chromatographic analysis. 2. Enantiomers—Separation.
I. Title.
QD117.C5A33 1999
543'.089—dc21   98-55565

1 3 5 7 9 8 6 4 2
Printed in the United States of America
on acid-free paper

This book is dedicated to the pioneers, authors, and column manufacturers in chiral separations, who have helped to move this field of science at a rapid pace.

PREFACE

This book has been planned for use by scientists interested in separations of isomeric compounds, especially those compounds that are chiral in nature—that is, are optically active. Information is provided on stereochemistry, separation methods (TLC, GLC, HPLC, and CE), and detection methods. Detailed discussion is given on HPLC methods. This includes derivatization, mobile-phase additives, ligand exchange, ion pairing, inclusion, and various chiral stationary phases (CSPs)—namely, brush-type, cellulosic, cyclodextrins, and protein-based. Some insight is provided into modern methods used in the design of CSPs, with special emphasis on target-directed designing methods. Detailed discussion includes method development and the choice of the right method and CSP. The importance of preparative separations for chiral compounds is discussed. Regulations and requirements are covered as they apply to development of chiral or enantiomeric compounds.

Stereoisomerism can result from a variety of sources aside from the single asymmetric carbon (chiral or stereogenic center). A molecule with a stereogenic axis can also be chiral. As a matter of fact, stereoisomerism can result from various sources (see chapter 3).

As discussed in chapter 1, enantiomers have identical physical properties except for the rotation of the plane of polarized light—that is, they have a different sign for optical rotation (plus or minus). Optically active molecules have attracted great attention because living systems are chiral. Proteins, nucleic acids, and polysaccharides possess chirally characteristic structures that are closely related to their functions. Because of chirality, living organisms usually show different biological responses to one of a pair of enantiomers (optical isomers) in drugs, pesticides, food, and waste compounds.

The importance of determining the stereoisomeric composition of chemical compounds, especially those of pharmaceutical significance, cannot be overemphasized, as covered in chapter 2. The differences in pharmacologic activity is exemplified by dextromethorphan, which is an over-the-counter antitussive (cough suppressant), whereas levomethorphan is a controlled narcotic. Less dramatic examples abound in that about 60% of the most prescribed drugs possess one or more asymmetric centers in the drug molecule. A fairly large percentage of the commonly used drugs are used in either racemic or diastereomeric forms. The differences in physiologic properties between the enantiomers of these drugs have not yet been examined in many cases, mainly because of the difficulties of obtaining both enantiomers in optically pure forms.

The primary focus of this book is on the methods used most commonly, namely, chromatographic methods:

- Thin-layer chromatography
- Gas chromatography
- High pressure liquid chromatography
- Supercritical fluid chromatography
- Capillary electrophoresis
- Membrane separations

Discussed at length are the first three methods; information is also provided on the emerging supercritical methods. High pressure liquid chromatography has been discussed extensively because it has greater versatility than the other chromatographic methods. The methods covered also include usage of formation of diastereomers where these reactions favor separation or simplify it in terms of utilization of materials. The last two methods listed above, which are not strictly chromatographic, provide interesting avenues for separations.

Discussion of the desirable features of CSPs and understanding chiral chromatography in chapters 6 and 7 should be of great help in method development, which is covered in chapter 8. Assistance with the most daunting task of method selection is covered at length in chapter 10, and this chapter also provides selected applications to help the readers solve their individual problems.

This book is based on an American Chemical Society short course organized by me and taught jointly with Professor William Pirkle and Dr. Christopher Welch. Somehow or other I got volunteered into writing this book as I bought into their inputs that their schedules would not permit this hard commitment. I will always cherish their contributions to this course and the camaraderie developed over the time we worked together. The book tries to fill the gap of the current textbooks in the field of chiral separations by chromatography, notwithstanding an excellent contribution to this field by Stig Allenmark in 1991.

I would like to thank Bill Pirkle and Chris Welch for their contribution to the ACS copyrighted course. Many thanks to my wife, Fay, for being continuously helpful in numerous ways.

Calabash, North Carolina S. A.
December 1999

CONTENTS

Chapter 1 An Overview 3
    Chirality and Biological Activity 3
    Chiral Separations 6
    Chiral Stationary Phases 8
    Modern CSP Design 9
    Preparative Separations 10
    Regulatory Perspectives 11

Chapter 2 Perspectives on Evaluation of Stereoisomers
          as Drug Candidates 14
    Marketing Status of Single Isomers 16
    Development of Single Enantiomers 16
    Pharmacokinetics 23
    Detoxification 26
    Stability 29
    Industrial Perspectives 30
    Regulatory Guidelines 33

Chapter 3 Review of Stereochemistry 38
    Brief History 39
    Stereoisomerism 40
    Classification 45
    Nomenclature 46
    Resolution 48

Enantiopurity   49
Determination of Absolute Configuration   52

Chapter 4   Separation Methods   58
Thin-Layer Chromatography   59
Gas Chromatography   63
High Pressure Liquid Chromatography   74
Supercritical Fluid Chromatography   76
Capillary Electrophoresis   77

Chapter 5   Detection Methods   81
Detectors for TLC   81
Detectors for GC   82
Detectors for HPLC   83
Specific Detectors for Chiral Compounds   85

Chapter 6   Desirable Features of Chiral Stationary Phases   92
Desirable Features   92
Types of CSPs   96
Comparison of Various CSPs   108

Chapter 7   Understanding Chiral Chromatography   111
Enantioselective Interactions   111
Chromatographic Enantioselectivity   114
Inclusion   120
Transition Metal Complexes   122
Separations on Protein Columns   122
Molecular Modeling   125

Chapter 8   Method Development   128
Stereoisomeric Interactions   128
Chromatographic Methods   130
Modes of Separation in HPLC   131
Chromatography of Diastereomeric Derivatives   132
Enantiomeric Resolution using Chiral Mobile-Phase Additives   132
Enantiomeric Resolution using Chiral Stationary Phases   135
Computerized Methods   140

Chapter 9   Preparative Separations   145
Displacement Chromatography   146
Elution Chromatography   147

Chapter 10 Method Selection and Selected Applications 168
    Where to Start? 168
    Study the Molecule 169
    Which Method to Use? 170
    High Pressure Liquid Chromatography 170
    Gas Chromatography 208
    Supercritical Fluid Chromatography (SFC) 212
    Capillary Electrophoresis (CE) 214
    Comparative Separations 216

Appendix 221
Index 229

Chiral Separations by
CHROMATOGRAPHY

1

An Overview

Molecules that relate to each other as an object and its nonsuperimposable mirror image are chiral (from the Greek word *cheir*, meaning hand: they are like a pair of hands). The chiral molecules are also called enantiomers. A pair of enantiomers is possible for all molecules containing a single asymmetric carbon atom (one with four different groups attached). The asymmetric carbon has also been called the chiral center or stereogenic center (see definitions, at the end of this chapter). The chiral molecules are stereoisomeric in nature—that is, they are isomeric molecules with identical constitution but a different spatial arrangement of atoms. The symmetry factor classifies stereoisomers as either chiral or achiral molecules.

Diastereoisomers, or more simply diastereomers, are basically stereoisomers that have more than one stereogenic center and are not enantiomers of each other. Although a molecule may have only one enantiomer, it can have several diastereomers or none at all. It is important to remember that two stereoisomers cannot be both enantiomers and diastereomers of each other simultaneously.

Stereoisomerism can result from a variety of sources aside from the single chiral carbon or chiral center already mentioned (see also chapter 3). The point to remember is that it is not necessary for a molecule to have a chiral carbon in order to exist in enantiomeric forms, but it is necessary for a molecule as a whole to be chiral.

Chirality and Biological Activity

Enantiomers have identical physical and chemical properties except for the rotation of the plane of polarized light—that is, they have a different sign for optical rotation (plus or minus). The molecules with (+) optical rotation are called dextrorotatory (e.g., d-amphetamine), and those with (−) rotation are called levorotatory (e.g., l-amphetamine). Optically active molecules have attracted great attention because living systems are chiral. Proteins, nucleic acids, and polysaccharides possess chiral characteristic structures that are closely related to their functions. Because of chirality, living organisms usually show different biological responses to one of a pair of enantiomers (optical isomers) in drugs, food, pesticides, and waste compounds. For example, depending on the taster, sodium L-(+) glutamate tastes good, whereas its mirror image, D-(−)glutamate, tastes bitter or flat. To obtain a better understanding of D and L, as well as R, S nomenclatures, the reader may refer to the Appendix of this chapter or the discussion in chapter 3.

A mixture consisting of equal amounts of enantiomers — that is, a racemate — is obtained experimentally by chemical reactions carried out in an achiral environment. To obtain an optically pure species, the separation of an enantiomeric mixture, or optical resolution, is necessary. Furthermore, an accurate assessment of the isomeric purity of a substance is critical since isomeric impurities may have unwanted toxicologic, pharmacologic, or other effects. These impurities may be carried through the synthesis and preferentially act at one or more steps, yielding an undesirable level of another impurity.[1] It is not uncommon for one isomer of a series to produce the desired effect, while another may be inactive or even produce undesired effects. Some examples of activity differences between chiral molecules are given in Table 1.1. It is noteworthy that d-amphetamine is a potent central nervous system stimulant, whereas the l-isomer has little, if any, effect. Furthermore, biological activity does not always favor the d-isomer — for example, see epineph-

Table 1.1 Biological Activities of Some Isomeric Compounds

Category/Name	Activity
Drugs	
Amphetamine	d-Isomer is a potent central nervous system stimulant, whereas the l-isomer has little, if any, effect.
Epinephrine	l-Isomer is 10 times more active as a vasoconstrictor than the d-isomer.
Propanolol	Racemic compound is used as a drug; however; only (S)-(−)-isomer has the desired adrenergic activity.
Food	
Asparagine	d-Asparagine tastes sweet, while l-enantiomer tastes bitter.
Carvone	(S)-(+)-Carvone smells like caraway, while (R)-(−)-carvone smells like spearmint.
Insecticide	
Bermethrine	The d-isomer is much more toxic than the l-isomer.
Vitamin	
Ascorbic acid	(+)-Isomer is a good antiascorbutic, while (−)-isomer has no such properties.

Source: Reproduced from S. Ahuja, *Chiral Separations: Applications and Technology*, American Chemical Society, Washington, D.C., 1997. Copyright American Chemical Society 1997.

rine. Large differences in activity between enantiomers point out the need to accurately assess isomeric purity of pharmaceutical, agricultural, food, and other chemical entities.

The importance of determining the stereoisomeric composition of chemical compounds, especially those of pharmaceutical importance, cannot be overemphasized.[2] The most often quoted dramatic, though moot, example in this area is thalidomide — the teratogenic activity that leads to deformation of children has been claimed to be due exclusively to the (S)-enantiomer. The differences in pharmacological activity are better exemplified by dextromethorphan, which is an over-the-counter antitussive (cough suppressant), whereas levomethorphan is a controlled narcotic. Less dramatic examples can be found in that about 60% of the top prescribed drugs possess one or more asymmetric centers in the drug molecule. A significant percentage of the commonly used drugs are used in either the racemic or the diastereomeric form. The differences in physiologic properties between the enantiomers of these drugs has not been fully examined in many cases because of the difficulties in obtaining both enantiomers in optically pure forms. The patient may be taking a useless, or even undesirable, enantiomer when ingesting a racemic mixture, because, as mentioned, the enantiomers can exhibit different biological properties.

The world market for enantiomeric drugs exceeded $96 billion in 1998 (Table 1.2) and is expected to surpass well over $100 billion in the year 2000. This increase is the result of our interest in finding drugs with greater therapeutic activity and low toxicity. It is not uncommon for one of the enantiomers to be significantly active or toxic.

To ensure the safety and pharmacological effects of currently used and newly developing drugs, it is important to isolate and examine the enantiomers separately. Furthermore, it is necessary to measure and control the stereochemical composition of drugs because of the potential for specific problems:

Table 1.2 Sales of Enantiopure Drugs

Type	Sales (billion $)
Antibiotics	23.2
Anticancer	7.6
Antiviral	6.2
Cardiovascular	21.1
Central nervous system	7.8
Gastrointestinal	1.4
Hematologic	6.2
Hormones	11.6
Respiratory	4.2
Miscellaneous	6.9
Total	96.2

Source: *Chem. Eng. News*, Oct. 1999.

1. During manufacture, problems of physical separations or preparative-scale separations may arise.
2. During quality control or regulatory analysis, questions of purity and stability predominate.
3. During pharmacologic studies of plasma disposition and drug efficacy, toxicity of metabolites is a primary concern.

Chiral Separations

A large number of approaches can be used to obtain chiral molecules. Since Louis Pasteur reported the first example of optical resolution in 1848, a very large number of compounds have been resolved, mainly by fractional crystallization of the diastereomeric salts. The following methods can be used for separations of chiral molecules:[3]

1. Separation of enantiomers by crystallization
   a. Crystal packing triage
   b. Conglomerates
   c. Preferential crystallization
   d. Preferential crystallization in the presence of additives
   e. Asymmetric transformation of racemates
2. Chemical separation of enantiomers via diastereomers
   a. Formation and separation of diastereomer-resolving agents
   b. Separation via complexes and inclusion compounds
   c. Asymmetric transformation of diastereomers
3. Enantiomeric enrichment
4. Kinetic resolution
5. Enzymatic resolution
6. Partition in heterogeneous solvent mixtures

Discussion on these methods can be found in some recent books.[1, 3–6] The primary focus of this book is on the methods used most commonly — that is, chromatographic methods — some of which can be related to partition in heterogeneous solvent mixtures. The following chromatographic methods are discussed:

- Thin-layer chromatography
- Gas chromatography
- High pressure liquid chromatography
- Supercritical fluid chromatography
- Capillary electrophoresis
- Membrane separations

The first three methods are discussed at length, and information is also provided on the emerging supercritical methods. High pressure liquid chromatography has been discussed extensively because it is more versatile than the other chromatographic methods. The methods also include using formation of diastereomers when these reactions either favor separation or simplify it in their use of materials.

The last two methods are not strictly chromatographic, but are included in this book to provide interesting avenues for separations.

Chromatographic methods are considered most useful for enantiomeric resolution for a number of reasons, the most important being the ease of separation. It is generally possible to find a chromatographic method that can provide separation in a matter of minutes. This comment does not imply that the knowledge base for identifying such a method already exists or is well categorized. It is meant to highlight the flexibility and efficiency that chromatographic methods offer. This book should provide a better understanding of the available methods so that the reader can design an intelligent strategy that would lead to fast selection of a reliable method for their particular separation.

A historical account of chiral separations by chromatography is given in Table 1.3. It may be noticed that direct resolution of enantiomers by gas chromatography was first reported in 1966, and resolution of enantiomers by liquid chromatography was first achieved via ligand exchange in 1971. Since then, great progress has been made in this field, leading to the development of a number of other approaches for separations, which has in some ways made HPLC the premier technique for separation of chiral compounds. More recently, capillary electrophoresis has offered some unique separations at the analytical scale because of the high efficiency offered by this technique. However, this technique is not very useful for preparative separations. In most cases, HPLC is the technique of choice for preparative separations; membrane separations and supercritical liquid chromatography may provide some other avenues.

Chromatographic methods can be direct or indirect. Indirect methods entail derivatization of a given enantiomeric mixture with a chiral reagent, leading to a pair of diastereomers that can be resolved by a given chromatographic method. In the indirect approach, the enantiomer may be converted into covalent, diastere-

Table 1.3 Historical Account of Chiral Separations by Chromatography

1939	Henderson and Rule: Resolution of racemic camphor derivatives
1952	Dalgliesh: Three-point rule in paper chromatography of amino acids
1966	Gil-Av[a] et al.:[a] Direct resolution of enantiomers by GC
1971	Davankov and Rogozhin: Chiral ligand exchange chromatography
1972	Wulff and Sarhan: Enzyme analogue polymers for chiral LC
1973	Hesse and Hagel: Cellulose triacetate used for chiral resolution
1973	Stewart and Doherty: Agarose-bonded bovine serum albumin (BSA) for chiral resolution
1974	Blaschke: Synthesis of chiral polymers
1975	Cram et al: Chromatography with chiral crown ethers
1979	Pirkle[a] and House:[a] Synthesis of first silica-bonded CSP
1979	Okamoto[a] et al.:[a] Synthesis of helical polymers for chiral LC
1982	Allenmark:[a] Use of agarose-bonded BSA in chiral LC
1983	Hermansson: Use of silica-bonded α_1-acid glycoprotein for chiral resolution
1984	Armstrong[a] and DeMond:[a] Use of silica-bonded cyclodextrins

a. For these contributions, see S. Ahuja, *Chiral Separations by Liquid Chromatography*, ACS Symposium Series #471, Washington, D.C., 1991.

omeric compounds by a reaction with a chiral reagent; and these diastereomers are typically separated on a routine achiral stationary phase. The direct methods use chiral stationary or mobile-phase systems and are used more frequently because of ease of operation; however, selection of the right combination still remains an intriguing process. This approach has been extensively investigated by many research scientists. Early successful results did not attract much interest; the technique remained relatively dormant, and little was done to develop this approach into generally applicable methods. In the last 30 years, systematic research was initiated for the design of chiral stationary phases functioning to separate enantiomers by gas chromatography. Molecular design and preparation of chiral systems for HPLC have been examined since then. More recently, efforts have been directed to finding new types of chiral stationary and mobile phases on the basis of stereochemical viewpoint, resulting in the technical evolution of modern liquid chromatography.

In the direct approach, several variations that can be tried have been classified into two groups:

1. The enantiomers or their derivatives are passed through a column containing a chiral stationary phase.
2. The derivatives are passed through an achiral column using a chiral solvent or, more commonly, a mobile phase that contains a chiral additive.

In either variation of the second case, one depends on differential, transient diastereomer formation between the solutes and the selector to effect the observed separation.

Of various chromatographic methods, HPLC is easily one of the most powerful separation techniques. As a result, resolution of enantiomers by HPLC is expected to advance rapidly as more efficient chiral stationary phases are found.

Whereas separation methods such as TLC, GC, and HPLC are discussed at length in this book to give the reader an option in selection of analytical methods and to develop a better understanding of separation processes, the greatest focus has been given to HPLC because it offers the greatest promise at present. CE, membrane separations, and SFC are providing some insight into some of the new approaches that are likely to be used in this field.

Chiral Stationary Phases

The chromatographic separation of enantiomers can be achieved by various methods; however, it is always necessary to use some kind of chiral discriminator or selector (see chapter 4). Two different types of selectors can be used: chiral additive in the mobile phase and chiral stationary phases (direct methods). Another possibility is precolumn derivatization of the sample with chiral reagents to produce diastereomeric molecules, which can be separated by the nonchiral chromatographic method (indirect method).

Direct methods are more popular where a chiral stationary or mobile-phase system in chromatography is used. This approach is being examined extensively by many research scientists today. Early successful results did not attract adequate interest; as a result, the technique remained relatively dormant, and little was done to develop it into a generally applicable method. Nearly three decades ago, systematic research was initiated for the design of chiral stationary phases that would be useful for separation of enantiomers by gas chromatography. Molecular design and preparation of chiral phase systems for HPLC have been examined in some depth since then. More recent efforts have been made to find new types of chiral stationary and mobile phases on the basis of the stereochemical viewpoint, with the resultant technical evolution of modern liquid chromatography.

The mechanism of separation is dependent on the given mode of separation used. Some discussion on the mechanism of separation is provided later in this book; however, it should be recognized that detailed mechanisms for chiral separations have not been worked out. The proposals made by certain scientists appear attractive, but vigorous differences prevail, so an attempt has been made not to highlight any single proposal.

Enantiomers can be resolved by the formation of a diastereomeric complex between the solute and a chiral molecule that is bound to the stationary phase. The stationary phase is called a chiral stationary phase or CSP, and the use of these phases is the fastest-growing area of chiral separations. A large number of chiral phases have become commercially available since the development of the first successful HPLC-CSP by William Pirkle in 1981.[7]

The separation of enantiomeric compounds on CSP is due to differences in energy between temporary diastereomeric complexes formed between the solute isomers and the CSP; the larger the difference, the greater the separation. The observed retention and efficiency of a CSP is the total of all the interactions between the solutes and the CSP, including achiral interactions. Since there are so many HPLC-CSPs available to the chromatographer, it is difficult to determine which is most suitable to solve a particular problem. This difficulty can be largely overcome by grouping the CSPs for chiral separations according to a common characteristic (see chapter 6).

Modern CSP Design

Some of the recent developments in CSP design have extended the scope of enantiomeric separations to a point unimagined a few years ago. The enantiomers of literally tens of thousands of compounds can now be separated chromatographically, often with considerable understanding of the separation process. Such understanding enhances our ability to design improved CSPs and to assign absolute configuration from observed elution orders.

Initial efforts to use convenient chiral adsorbents (cellulose, starch, wool) usually met with little success. Columns containing swollen triacetylated microcrystalline cellulose have been used in both analytical and preparative modes. Although there is little detailed understanding as to how and why this CSP works, it is thought that the laminar nature of the swollen crystals offers chiral cavities into which enantiomers must intercalate. Cyclodextrin CSPs also require intercalation into the hole or cavity of the chiral cyclodextrin.

The mode of action of various cellulose derivatives coated onto diphenyl-silanized silica is poorly understood. Columns made from acylated, benzoylated, cinnamoylated, phenyl carbamoylated, and benzylated silicas are commercially available. While these columns afford a number of interesting enantiomeric separations, there is no clear understanding as to what will resolve on which column or in what order the enantiomers will elute. A variety of polysaccharides have been investigated; however, no clear pattern of performance is evident.

Protein columns — columns packed with silica-bound proteins — show a fairly extensive range of separations, although the nature of chiral recognition processes employed is still not very clear. The innate complexity of these biopolymers baffles one's ability to deduce the details of the operative chiral recognition processes. Moreover, it is not easy to alter or fine-tune the structure of a CSP to enhance selectivity.

The complexity of separations on polymeric CSPs stems from the analyte perceiving the CSP as a chiral array of subunits (monomers), which may themselves be chiral. Unless the structure of the array is known, it can be difficult to delineate a chiral recognition mechanism.. The abundance of closely spaced potential interaction sites makes the mechanistic understanding more difficult. This is also likely to be the case for the synthetically designed polymeric CSPs.

The understanding of chiral recognition and the design of rational stationary phases must have at least a two-pronged approach: first, consider chiral recognition as it might occur between small chiral molecules in solution, and then study the separation of polymeric separations in smaller arrays of monomers. Details of the rational design of chiral stationary phase are given in chapter 7.

Preparative Separations

Enantiomeric compounds can be prepared by a number of ways:

- Stereoselective synthesis
- Enzyme catalysis
- Separation methods

Enantioselective synthesis usually requires a chiral starting material as a building block or as an auxiliary, along with elaborate synthetic steps that would lead to production of only one enantiomer. Catalytic procedures in which chiral information is transferred from a chiral catalyst to the prochiral substrate or in which one enantiomer is preferably transformed — for example, enzyme catalysis — are also used at times. The degree of difficulty is rather high and can generally be related to the desired degree of optical or enantiomeric purity. (It should be noted that "optical purity" and "enantiomeric purity" are used interchangeably in this book; however, the latter term is preferable.)

Large-scale preparative liquid chromatographic systems are already available as process units for isolating and purifying chemicals and natural products. Chiral separations by HPLC are ideally suited for large-scale preparation of optical isomers. For large-scale separations and in consideration of the cost of plant-scale resolution processes, the sorption methods offer substantial increases in efficiency over recrystallization techniques.

For the separation methods, the material is prepared as the racemate by a reaction sequence, which generally presents a much lower degree of difficulty than for the corresponding optically active forms, and then the enantiomers are separated indirectly by forming diastereomers or directly by chromatography on a chiral stationary phase. Although the separation using the formation of diastereomers is still often used, especially for the compounds bearing acid or base functions, the contribution of direct separation of enantiomers by chiral stationary phases is rapidly increasing. The main reasons for the increased usage are that both enantiomers can be obtained for pharmacologic or toxicologic evaluations, a high degree of optical purity of the isolated enantiomers is required, the methods are rapid and achievable without great difficulty, and applications are possible to a broad variety of compounds that cannot be derivatized easily, such as hydrocarbons, those compounds that racemize easily, and those compounds that have helical or propeller-type chirality.

These methods are discussed at length in this book (see chapter 9). The drawbacks of these methods may include the high cost of the stationary phase, the high dilution factor, the consumption of a large amount of the mobile phase, and difficulties associated with recycling the mobile phase. The obstacles to scaleup can be largely overcome because of recent improvements in chromatographic techniques and the development of relatively cheap chiral stationary phases for preparative purposes. Chromatography is now regarded as technically and economically attractive for the preparation of high-value compounds and of optical isomers that are accessible only with great difficulty. Kilogram quantities of optically pure isomers are already being produced using this technology, and this approach offers a very useful method for the manufacture of chiral drugs that are very potent and cannot be easily prepared by other methods. Other preparative techniques such as simulated moving-bed chromatography and membrane separations are likely to further enhance the value of preparative separations.

Regulatory Perspectives

Some information is available on regulatory practices in Canada, the European Economic Community, Japan, and the United States in documents that are at various stages of development. All of the requirements are based on similar scientific backgrounds and mainly target pharmaceutical compounds. These requirements have evolved after discussions between various authorities within the industry at various international meetings. As a result, the texts are similar to each other in content (see chapter 2). The U.S. Food and Drug Administration perspective is summarized here, to give the reader a bird's-eye view of what is likely to be required by various regulatory agencies.

1. All chiral centers should be identified
2. The enantiomeric ratio should be defined for any admixture other than 50:50
3. Proof of structure should consider stereochemistry
4. Enantiomers may be considered impurities
5. Absolute configuration is necessary for an optically pure drug
6. Marketing an optical isomer requires a new drug application (NDA)
7. An investigational drug (IND) application is required for clinical testing
8. Justification of the racemate or any of the optically active forms must be made with the appropriate data
9. Pharmacokinetic behavior of the enantiomers should be investigated

This book deals at length with regulatory issues and how they are to be addressed from the standpoint of method development and technology.

Appendix: Definitions of Some Common Terms

Asymmetric Carbon Atom A carbon atom with four different substituents.

Chiral Molecule A molecule with at least one pair of enantiomers.

Chiral Center In a tetrahedral or trigonal pyramidal structure, the atom to which four different substituents are attached and to which descriptor R or S can be assigned (see below). Also see Stereogenic Center.

Configuration Spatial arrangement about a chiral atom.

Conformation The spatial array of atoms in a molecule of given constitution and configuration. Conformation of such molecules can be changed by (rapid) rotation around single bonds without, in general, affecting the constitution and configuration.

d or l Rotation Refers to dextro (right) or levo (left) rotation of the plane polarized light.

D- or L- Convention Assignment of configuration around a chiral atom by comparison to D-(+)-glyceraldehyde. The D- or L- designation does not correspond to d or l optical rotation.

Diastereoisomer or Diastereomers Optical isomers that are not related as object and its mirror image. More than one stereocenter is present.

Distomer Enantiomer with less desired biological activity.

Enantiomers Stereoisomers that are related as nonsuperimposable mirror images.

Epimers Diastereomers that differ in absolute configuration at one of two or more stereogenic centers.

Eudismic Ratio Ratio of the activity of the active enantiomer to that of the less active enantiomer.

Eutomer Enantiomer with desired biological activity.

Optical Rotation The angle of rotation of plane of polarized light by a chiral molecule.

R or S Convention Cahn-Ingold-Prelog system (see chapter 3).

Stereogenic Center A focus of stereoisomerism in a molecule such that one interchange of two substituents on an atom in such a molecule leads to a stereoisomer. Other terms used are stereogenic element, axis, or plane.

Stereoisomers Molecules with the same constitution that differ with respect to spatial arrangement of certain atoms or groups.

REFERENCES

1. S. Ahuja, *Chiral Separations: Applications and Technology.* American Chemical Society: Washington, DC, 1997.
2. S. Ahuja, paper presented at the First International Symposium on Separation of Chiral Molecules, Paris, 1988.
3. E. L. Eliel and S. H. Wilen, *Stereochemistry of Organic Compounds.* Wiley: New York, 1994.
4. I. W. Wainer, *Drug Stereochemistry.* Marcel Dekker: New York, 1993.
5. S. Allenmark, *Chromatographic Enantioseparation.* Ellis Horwood: New York, 1991.
6. S. Ahuja, *Chiral Separations by Liquid Chromatography.* American Chemical Society: Washington, DC, 1991.
7. W. H. Pirkle, J. M. Finn, J. L. Schreiner, and B. C. Hamper, *J. Am. Chem. Soc.*, 103, 3964 (1981).

2

Perspectives on Evaluation of Stereoisomers as Drug Candidates

A great deal of progress has been made in the area of regulations and requirements for chiral compounds since the first international symposium on chiral separations was held in Paris in 1988.[1] The magnitude of scientific exchange in this area since then indicates the importance of determining the stereoisomeric composition of chemical compounds, especially those of pharmaceutical importance.[2-10]

A number of examples have been provided in Table 1.1; these clearly show that stereoisomers can have different pharmacologic activities. (Remember that dextromethorphan is a well-known over-the-counter antitussive, whereas levomethorphan, its stereoisomer, is a controlled narcotic.) Similarly, some researchers speculate that the teratogenic activity of thalidomide may reside exclusively in the (S)-enantiomer.[11] This conclusion remains controversial because racemization was seemingly overlooked.[12]

Thalidomide (**2.1**) has been found to racemize with relative ease (residual enantiomeric excess of two-thirds after 5 days in DMF or 80% DMF at room temperature).

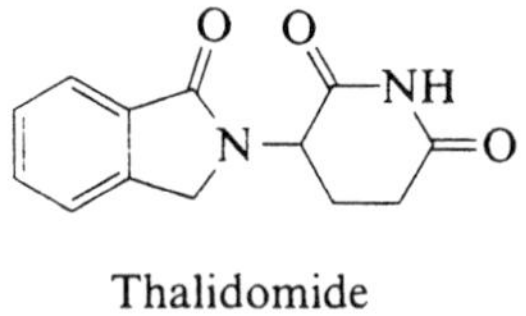

Thalidomide

2.1

Studies under biomimetic conditions yielded a racemization half-life of 2.5 hours in phosphate buffer at pH 7.4 and 37°C, with serum albumin markedly accelerating the reaction. In fact, thalidomide isolated from the plasma of rabbits 2 hours after intravenous injection was completely racemized. Facile racemization has also been demonstrated by two analogues of thalidomide, 2-phthalimidinoglutarimide and 2-phthalimidoadipinimide. The former compound, for example, racemized in marmosets with a $t_{1/2}$ of about 3 hours. Such results have generated a lively debate; the consensus is that it is practically impossible to demonstrate stereoselectivity in any in vivo biological effect of thalidomide. Only in vitro tests of short duration can be expected to yield reliable indications.

These observations emphasize the need for careful investigations because a large number of the most frequently prescribed drugs in the United States and more than 50% of the top 200 drugs possess one or more asymmetric centers in the drug molecule.[13] About half of the drugs listed in the *U.S. Pharmacopoeial Dictionary of Drug Names* contain at least one asymmetric center, and a sizable number of them have been used in racemic or diastereomeric forms.[14]

Remember that the differences in the biological properties between enantiomers of these racemic drugs have not yet been determined in many cases, probably because of the difficulties of obtaining both enantiomers in optically pure forms. Some enantiomers may exhibit potentially different pharmacologic activities, and the patient may be taking a useless or even undesirable enantiomer when ingesting a racemic mixture. To ensure the safety and desired effect of currently used and newly developing drugs, it is important to isolate and examine both enantiomers separately. Furthermore, as already mentioned, the stereochemical composition of drugs must be measured and controlled during manufacture, when problems of preparative-scale separations may be involved; during quality control (or regulatory analysis), when analytical questions of purity and stability predominate; and during pharmacologic studies of plasma disposition and drug efficacy, when ultratrace methods may be required.[4]

Based on the sales of chiral drugs (see Table 1.2) and the rapid growth of chiral drugs, the market size for chiral compounds is likely to well surpass $100 billion in the year 2000. Accurate assessment of the isomeric purity of substances is critical, since isomeric impurities may have unwanted toxicologic, pharmacologic, or other effects. Such impurities may be carried through a synthesis and react preferentially at one or more steps, yielding an undesirable level of another impurity. Frequently, one isomer of a series produces a desired effect, while another is inactive or even produces some undesired effect. Large differences in activity between stereoisomers point out the need to accurately assess isomeric purity of pharmaceuticals. Often these differences exist between enantiomers, the stereoisomers most difficult to separate.

Discussed below are the marketing status of single enantiomers and the importance of carefully evaluating stereoisomers to avoid untoward biological effects and instability. The perspectives of pharmaceutical manufacturers and regulatory bodies are provided, followed by detailed discussion of guidelines from the U.S. Food and Drug Administration (FDA) and requirements from the regulatory standpoint for absolute configuration and enantiomeric impurities.

Marketing Status of Single Isomers

The results of a recent study[15] indicate that the use of single-isomer chiral drugs increased in the 1990s. The main reason for this increase appears to be a greater emphasis on the development of synthetic single-isomer chiral drugs. This is exemplified by the introduction of a number of angiotensin-converting enzyme inhibitors during this period. Two results of this survey appear to be particularly striking: (1) the overall reduction in the number of compounds of natural and semisynthetic origin in the 1990s as compared to the 1980s and (2) the large increase in the number of synthetic single-isomer chiral drugs.

This has led to reconsideration of the classification of the origin of the drugs studied in this survey. The majority of previous surveys of this type gave no clear indication of the classification used other than natural, semisynthetic, or synthetic origin. In this survey, designations were based on the origin of the category considered to be most appropriate in terms of the present-day source of the drug used in the production of formulated products; for example, ascorbic acid and caffeine were both defined as being of synthetic origin. In previous surveys, these two compounds (and others included in this survey) may well have been classified as being of natural origin. This may have resulted in some minor differences in the reported results in comparison to previous studies; however, it can be concluded that the presented results are a true reflection of the changes that have occurred in the use of single-isomer chiral drugs over the years.

Development of Single Enantiomers

It is well known that enantiomers can have completely different pharmacologic, biochemical, or toxicologic profiles.[15] The enantiomers of promethazine are almost equivalent in terms of their antihistaminic properties and toxicity, yet the β-blocking activity of (−)-S-propanolol is considerably greater than that of (+)-R-propanolol.[16] With the intravenous anesthetic ketamine, which is routinely administered as the racemate, reports indicate that the (+)-S-isomer is superior to the (−)-R-isomer in terms of the provision of adequate anesthesia. Additionally, the (−)-R-isomer of ketamine has been shown to be the major cause of the postoperative side effects (hallucinations and other transient psychotic sequelae) observed with the use of racemic ketamine.

Development of single-isomer drugs can result in decreased toxicity in comparison to their racemates, as exemplified by D-penicillamine and L-dopa.[17] As both of these compounds are relatively old, the decision to use single isomers most likely relates to the technology available at the time of their development, since both were based on amino acid chemistry. Hence, this supports the argument that with current technology, single-isomer drugs should become the norm rather than the exception. A clear suggestion evolves out of this discussion that stereochemical differences in pharmacology and toxicology should be studied early in drug development so that at that stage a rational decision can be made on the material to be finally developed and marketed.

A potential commercial problem can arise if a single isomer of a previously marketed racemate could be patented by a company other than the compound's originator. In this situation, obviously the originator's competitive position in the marketplace would be affected. The significance of chirality in the assessment of human and veterinary marketing authorizations needs to be addressed. It may be emphasized that in all cases, stereoisomers are uniquely different compounds rather than just different forms of the same compound.

Enantiomers of chiral compounds can differ widely in biological activity, qualitatively as well as quantitatively.[18] Nevertheless, most of the pharmacologic data available to date on chiral drugs are obtained from experiments with racemates, which assume that the biological activity generally resides in one of the enantiomers. With the advances made in stereospecific synthesis and stereoselective analysis of drugs, pharmacologists can now explore the steric aspects of drug action. Unfortunately, the degree of resolution is seldom specified in published works on stereoselectivity of drugs. Discussed below are examples of derivatives of phenylethylamine that act with adrenergic mechanisms.

Compounds with One Chiral Center

Terbutaline (**2.2**) is a β_2-selective adrenoceptor agonist with one chiral center.

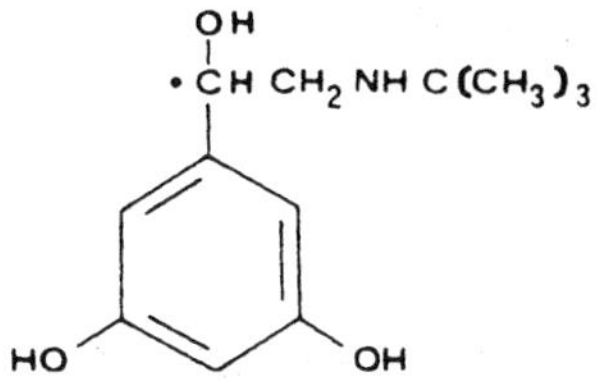

$$\text{•CH(OH)—CH}_2\text{—NH—C(CH}_3\text{)}_3$$

(attached to 3,5-dihydroxyphenyl ring)

2.2

The (−)-enantiomer of this compound is a potent relaxant of tracheal smooth muscle, while the (+)-enantiomer is approximately 3,000 times less active in this respect (Table 2.1). To achieve this result, the distomer (less active isomer) must not be contaminated with more than 0.03% of the eutomer (active isomer). The effect of both enantiomers is blocked by racemic propranolol, but the distomer markedly less than the eutomer. This may indicate an unspecific relaxing power of the distomer at the very high concentrations in question. Alternatively, an atypical β-adrenoceptor is involved. It seems that the validity of the eudismic ratio obtained experimentally is limited by both the enantiomeric purity of the drug and problems with specificity at very high drug concentrations. No interaction was found between the two enantiomers of terbutaline on tracheal smooth muscle, even when the distereomer is at hundredfold excess (Table 2.1).

Detailed functional studies in vitro with the enantiomers of terbutaline on tracheal (mainly β_2-adrenoceptors), skeletal (soleus, β_2), and cardiac (papillary, β_1)

Table 2.1 Steric Aspects of Agonism (pD_2) at β-Adrenoceptors in Guinea Pig Trachea

Compound	Potency	Eudismic ratio
(−)-Terbutaline	7.28	3,300
(+)-Terbutaline	3.76	
(−)-Terbutaline with 5 mmol/liter of (+)-terbutaline	7.78	
(−)-Clenbuterol	8.54	>10,000
(+)-Clenbuterol	<5.5	

Source: Reproduced with permission from B. Waldeck, *Chirality*, 5, 355 (1993). Copyright 1993 Wiley-Liss.

muscle show that (−)-terbutaline is a $β_2$-selective agonist of its own and that the (+)-enantiomer is several thousandfold weaker at both receptor subtypes.

Racemates can have multiple effects. One of the best known examples is the well-known $β_1$-selective adrenoceptor agonist dobutamine. This compound, widely used as a tool for characterizing receptors, consists of one β-agonist, the (+)-enantiomer, and a partial α-agonist, the (−)-enantiomer. Another example of dual properties of a chiral adrenoceptor ligand is the antagonist amosulalol (**2.3**).

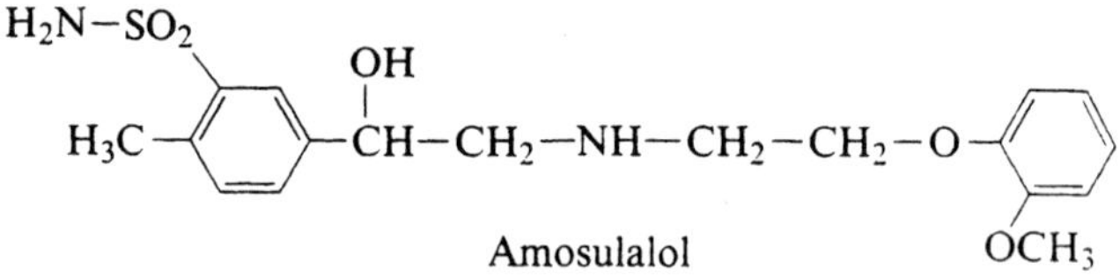

$$H_2N-SO_2\text{-}H_3C\text{-}\underset{\text{Amosulalol}}{\text{C}_6H_3}\text{-}\underset{\text{OH}}{\overset{|}{CH}}-CH_2-NH-CH_2-CH_2-O-C_6H_4-OCH_3$$

Amosulalol

2.3

The (−)-enantiomer of this derivative of phenylethanolamine inhibits β-adrenoceptors unselectively and, in the same concentration range, $α_1$-adrenoceptors; however, a 100-fold higher concentration is required to inhibit $α_2$-adrenoceptors (Table 2.2). The (+)-enantiomer, however, is a potent and highly $α_1$-selective antagonist with a much lower affinity for β-adrenoceptors. Interestingly, the eudismic ratio (eutomer/distomer) is much higher for interaction with β-receptors than for α-receptors; for β-effects, the (−)-isomer is the eutomer, while the condition is reversed for α-effects.

Removal of the β-hydroxyl group of amosulalol results in a marked drop in its affinity for β-receptors while the affinity for α-receptors is unchanged ($α_1$) or even increased ($α_2$). This finding does not fully comply with the Easson-Stedman hypothesis, which postulates a three-point interaction of the drug with the receptor. When adopted for phenylethanolamines, this hypothesis says that if the β-hydroxyl

Table 2.2 Effect of Amosulalol and Its Desoxy Derivative on Receptors' pA_2 Values

Receptor	(−)	(+)	Desoxy	Eudismic ratio
β_1, Rat right atrium	7.71	6.03	6.19	48
β_2, Guinea pig trachea	7.38	5.71	5.76	47
α_1, Rabbit aorta	7.17	8.31	8.14	14
α_2, Rat vas deferens	4.92	5.36	6.05	3

Source: Reproduced with permission from B. Waldeck, *Chirality*, **5**, 355 (1993). Copyright 1993 Wiley-Liss.

group is in the "wrong" position (S-configuration), the potency of the compound equals that of the desoxy derivative. Since both adrenaline and noradrenaline appear to comply with the Easson-Stedman hypothesis for both receptor types, presumably it is the phenoxyethylamine rather than the phenylethanolamine part of amosulalol that interacts with the α-adrenoceptors. Thus α- and β-adrenoceptors may interact with different parts of the same molecule. This suggestion is supported also by the "anti-Pfeiffer" behavior of amosulalol. Pfeiffer's rule says in principle that the eudismic ratio for enantiomer pairs increases with increased affinity of the eutomer. Amosulalol has a high affinity and a low eudismic ratio when it interacts with β-adrenoceptors. This anomaly can be explained by proposing that the phenoxyethylamine part of the molecule interacts with the α-adrenoceptor, in which case the β-hydroxyl group will have a more remote position. Similar results and a similar interpretation have been given for the enantiomers of carvedilol, another α- and β-adrenoceptor antagonist.

Compounds with Two Chiral Centers

With two chiral centers, four different stereoisomers (two diastereomer pairs) are possible. Although many drugs have two chiral centers, there are few examples of a complete pharmacologic examination of the different enantiomers. Among the N-substituted phenylethanolamines, there are three structurally related adrenoceptor ligands, two agonists, and one antagonist, whose individual stereoisomers have been investigated. These compounds are the β_2-selective agonist formoterol (**2.4**), a *p*-trifluoromethyl anilide derivative (PTFMA) with β-agonistic properties, and labetalol (**2.5**), which blocks α- and β-adrenoceptors.

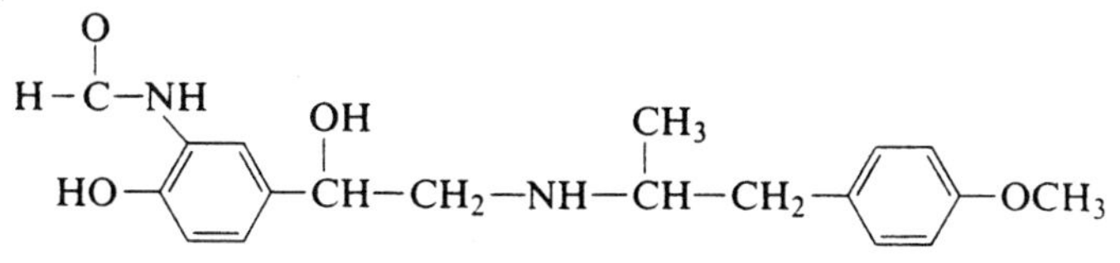

Formoterol

2.4

$$H_2N-CO\text{-}\underset{HO}{\bigcirc}\text{-}\underset{OH}{CH}\text{-}CH_2\text{-}NH\text{-}\underset{CH_3}{CH}\text{-}CH_2\text{-}CH_2\text{-}\bigcirc$$

Labetalol

2.5

For the enantiomers of formoterol and their diastereomers, the order of potency is (R;R) >> (R;S) = (S;R) > (S;S) with respect to relaxation of airway smooth muscle. Comparable results were obtained with PTFMA for the activation of adenylate cyclase in turkey erythrocytes and for the stereoisomers of labetalol with respect to inhibition of β_1-adrenoceptors in guinea pig heart.

The (R;R)-enantiomer is the most potent in both compounds, followed by the (R;S)-diastereomer. A common feature is also the relatively high eudismic ratio (R;R)/(S;S) and the low ratio of (R;S)/(S;R). This suggests that the configuration of the nitrogen substituent is critical for the interaction of the phenylethanolamine moiety of the molecule with the β-adrenoceptor.

There appear to be quite different steric requirements for the interaction of labetalol with the α-adrenoceptor. In this respect, the (S;R)-isomer is most potent, and the eudismic ratio (S;R)/(R;S) is about 50. There is no clear difference in potency between the (R;R)- and (S;S)-enantiomers, which is in sharp contrast to the condition for the interaction with β-adrenoceptors. It appears that α- and β-adrenoceptors interact with different parts of the molecule.

From these examples of structurally related compounds, it appears that the (R)-configuration at the carbon atom carrying the hydroxyl group is essential for interaction with β-adrenoceptors. The influence of the configuration at the carbon atom attached to the nitrogen is variable, but in no case is the (R;S)-isomer more potent than the (R;R)-isomer. It should be noted, however, that between the (R;R)- and (R;S)-diastereomers, there may be not only potency differences but also differences in selectivity for β-adrenoceptors in different tissues.

Enantiomeric Purity and Biological Consequences

The reported eudismic ratio may differ from one laboratory to another. For example, the eudismic ratio for the relaxant effect of salbutamol on guinea pig tracheal smooth muscle has been reported at 70 to 300. Part of this variation may be due to traces of the more active isomer in the less active one, a factor that becomes more important with the higher true eudismic ratio, as was pointed out by Barlow et al.[19] more than 20 years ago. From those observations, it follows that the higher the observed eudismic ratio, the more uncertain the true ratio, since minute traces of the eutomer in the distomer are difficult to detect by analytical methods.

The degree of resolution is rarely specified in the pharmacologic literature, and when it is, there is often only a note on the optical rotation. This measure of enantiomeric purity is far from reliable when the enantiomeric contamination is below a few percentages. Since eudismic ratios observed in biological experiments often fall in the range of 100 to 1,000, more sensitive analytical methods are necessary. When the degree of enantiomeric purity is uncertain, conclusions regarding chiral aspects of structure-activity relationships, including evaluation of Pfeiffer's rule, should be made with care. Subtle differences between diastereomers, for example, must be interpreted with caution when the enantiomeric purity is not known. This is illustrated below with data on formoterol.

In the first report on the stereoisomers of formoterol, the highest potency ratio obtained for the relaxation of tracheal smooth muscle was about 14, far smaller than what might be expected and probably a result of incomplete resolution.[18] When the contamination with the (R;R)-isomer was reduced from 1% to 0.1%, the eudismic ratio (R;R)/(S;S) increased from 50 to 850, and the order of potency between the (S;R)- and (S;S)-isomers was reversed. At the same time, the potency difference between the (R;S)- and (S;R)-enantiomers disappeared. A further increase in purity might reduce the potency of the (S;S)-enantiomer even more, but the potency of the (S;R)-enantiomer appears to have approached its true value.

In pharmacologic characterization of chiral compounds, it is essential to find out whether the distomer may have adverse effects. In experiments addressing this question, very high doses are usually employed. If the distomer in this case is contaminated with the eutomer, the effects observed may be due to the traces of the eutomer or may be the result of an interaction between the enantiomers at a fixed ratio different from the racemate rather than the effect of the distomer, per se. This possibility must be considered in safety assessment.

A controversy exists concerning the therapeutic use of nonsteroidal anti-inflammatory drugs (NSAIDs) as to whether these drugs should be produced, marketed, and used clinically as the racemate or as a single enantiomer.[20] Traditionally, the therapeutic and major toxic effects of NSAIDs have been attributed to the ability of these drugs to inhibit the synthesis of stable prostaglandins through the direct inhibition of prostaglandin H synthetase, which serves both as a cyclooxygenase and as a peroxidase. These properties have been largely determined from the results of in vitro studies and in turn have shown a marked stereoselectivity for chiral NSAIDs, in favor of the (S)-enantiomers (30 to 100 times), as opposed to their R antipodes. Therefore, it has been suggested that the (R)-enantiomers of NSAIDs are unnecessary impurities or isomeric ballast and that a stereochemically pure enantiomer is superior to its respective racemate.

Malmberg and Yaksh[21] have suggested that the powerful anti-inflammatory effects of NSAIDs have most likely diverted attention away from many other properties of therapeutic relevance. For example, indomethacin is twice as potent in the inhibition of cyclic AMP-dependent protein kinase than in the inhibition of the prostaglandin H synthetase. Subsequently, many other non-prostaglandin-dependent mechanisms, including uncoupling of oxidative phosphorylation, inhibition of renal anion transport, changes in neutrophil and leukocyte function, and interruption of signal transduction through G proteins, have been identified as therapeutically relevant.

The most significant adverse effects of NSAID therapy occur in the gastrointestinal (GI) system. The GI effects of NSAIDs have been presented predominantly as ulceration of the stomach and duodenum. There is a growing body of evidence, however, that more-distal (small-intestinal) damage may be more widespread, persistent, and serious than previously thought. In addition, it has been demonstrated that NSAIDs cause increased small-intestinal permeability at the level of the mucosal tight junction, and that this may lead to intestinal inflammation. This, in turn, has been implicated in the genesis of the more serious intestinal sequelae. The permeability of tight junctions may be partly regulated by prostaglandins. Therefore, one can expect NSAIDs to cause increased intestinal permeability. Further, it may be reasonable to expect that the effect on intestinal permeability is almost exclusively related to the (S)-enantiomers.

Although etodolac is used clinically as a racemate, its prostaglandin synthetase inhibition has been attributed almost exclusively to the (S)-enantiomer. Due to metabolic inversion to the active (S)-enantiomer, the intrinsic effect of (R)-enantiomer is difficult to examine for many chiral NSAIDs; however, because of the fixed nature of its asymmetric center, the (R)-enantiomer of etodolac does not undergo chiral inversion. The unexpected increase in the urinary excretion of Cr-EDTA in rats, following (R)-etodolac administration, is direct evidence for its intrinsic activity in increasing small-intestinal permeability. This suggests that inhibition of prostaglandin synthetase may not be the only mechanism through which etodolac affects the GI tract. Indeed, (R)-etodolac is only one order of magnitude less potent in increasing intestinal permeability than the (S)-enantiomer, which contrasts with the up to 150-fold difference in potency observed in vitro against prostaglandin synthetase. Further, the increase in urinary excretion of Cr-EDTA by the (S)-enantiomer (6 mg/kg) appears to be equal to, if not greater than, that of the racemate (12 mg/kg containing 6 mg/kg (S)- and 6 mg/kg (R)-etodolac). This may suggest that despite the apparent activity of both enantiomers, their presence in the racemate may exert an ameliorating effect on the extent of intestinal permeability observed. The discrepancy between the effects of individual enantiomers and their racemate may result, in part, from differences in their physicochemical properties, which in turn may alter their biodistribution and even drug-receptor interaction. It has been shown that the enantiomers of ibuprofen have substantially different solubility parameters than racemic ibuprofen. In addition, the octanol/water partition coefficient of etodolac enantiomers has been shown to be affected by the presence of the antipode.

The data suggest that mechanisms other than peripheral prostaglandin synthetase inhibition may be responsible for the effects of etodolac on intestinal permeability. Further, the data demonstrating the activity of the (R)-enantiomer in increasing intestinal permeability are congruent with the observations of other investigators who have shown that the (R)-enantiomers of ibuprofen and flurbiprofen produce both analgesia and changes in polymorphonuclear lymphocyte function. From the therapeutic perspective, the decision to use either the racemate or stereochemically pure enantiomer must be based on examination of the relevant and possibly diverse mechanisms of action rather than on the single most obvious effect.

Pharmacokinetics

Pharmacokinetic studies are generally undertaken to define as accurately as possible the rates of absorption, metabolism, and excretion of a drug and its metabolites. The importance of these studies is exemplified by pirprofen, a member of the 2-arylpropionic acid (2-APA) class of nonsteroidal anti-inflammatory drugs (NSAIDs). The drug has a chiral center and was marketed as the racemate prior to its removal from the pharmaceutical market because of several reported cases of hepatotoxicity.[22] Similar to other chiral NSAIDs, the (S)-enantiomer of pirprofen possesses much more pharmacologic activity than its R antipode.

Stereoselective pharmacokinetics have been observed for several chiral NSAIDs. Furthermore, the degree of stereoselectivity is both drug- and species-specific. Examples include ketoprofen and flurbiprofen, which display a considerable degree of stereoselectivity in their pharmacokinetics in the rat, yet show little enantioselective disposition in humans. On the other hand, both humans and rats share similar patterns and degrees of stereoselectivity in the pharmacokinetics of etodolac, ketorolac, and fenoprofen.

For an important class of chiral NSAIDs, the 2-APA derivatives, part of the interspecies variation in the disposition of the enantiomers can be explained on the basis of the abilities of the species to bioinvert the (S)-enantiomer to its antipode. This pathway of metabolism may occur both systemically and presystemically. Therefore, the route of administration and the properties of the dosage form may influence the pharmacokinetics of 2-APA NSAIDs that undergo inversion.

The enantioselective disposition of pirprofen has been studied in 11 healthy human volunteers. Although pharmacokinetic indices were not reported, it appeared that after oral doses, the plasma concentrations of (S)-pirprofen were higher than those of the respective (R)-enantiomer. Similarly, in a preliminary study involving three female rats dosed orally, the areas under the plasma concentrations time curve (AUC) of (S)-pirprofen were 22% higher than those of the (R)-enantiomer.

The pharmacokinetics of the enantiomers of the NSAID drug pirprofen were studied in male Sprague-Dawley rats after oral and intravenous (IV) doses of the racemate. No significant differences were detected between the enantiomers after oral or IV dosing in t, Vd, or ΣXu. However, the R:S area under the plasma concentration ratio after oral doses (0.92 ± 0.13) was slightly but significantly lower than matching IV doses (1.05 ± 0.036). The absolute bioavailability of the active (S)-enantiomer (78.5%) after oral doses was higher than the inactive (R)-enantiomer (69.3%). The plasma protein binding of both enantiomers was saturable over a fivefold range of plasma concentrations. At higher plasma concentrations, the (S)-enantiomer was less bound than the (R)-enantiomer. In an in vitro experiment using everted rat jejunum, no chiral inversion was discernible. The dependency of the AUC ratio of the enantiomers on the route of administration may be due to stereoselective first-pass metabolism.

Deuterium labeling techniques and stereoselective GC/MS methodology have been employed to investigate the mechanism by which (R)-ibuprofen undergoes metabolic chiral inversion in the rat in vivo.[23] Following oral administration of a mixture of (R)-ibuprofen (7.5 mg kg^{-1}) and R-[ring-^{2}H$_4$; 2-^{2}H]ibuprofen (R-[^{2}H$_5$]ibuprofen)(7.5 mg kg^{-1}) in male Sprague-Dawley rats, the enantiomeric composition and deuterium excess of the drug were determined in serial plasma samples and in pooled urine collected over 10 hours. The results demonstrate that:

1. (R)-ibuprofen undergoes extensive inversion of configuration to its S antipode in the rat.
2. Chiral inversion of R-[^{2}H$_5$]ibuprofen yields S-[^{2}H$_4$]ibuprofen in a process that involves quantitative loss of the deuterium atom present originally at C-2.
3. Labeling of (R)-ibuprofen with deuterium at C-2 does not introduce a measurable kinetic deuterium isotope effect on the chiral inversion reaction.
4. Metabolism of R-[^{2}H$_5$]ibuprofen leads to the appearance in plasma and urine of molecules of (R)-ibuprofen labeled with four atoms of deuterium.

On the basis of these findings, a mechanism is proposed for the chiral inversion reaction that invokes the stereoselective formation of the coenzyme A thioester of (R)-ibuprofen as a key metabolite; conversion of this species to the corresponding enolate tautomer affords a symmetrical intermediate through which racemization of ibuprofen occurs in vivo.

The influence of aging on the pharmacokinetics and tissue distribution of (R)- and (S)-propranolol was studied in 3-, 12-, and 24-month-old rats. After both IV and oral administration of racemic propranolol, the plasma concentrations were higher for the (R)-enantiomer than for the (S)-enantiomer.[24] For the tissue concentrations, the reverse was true. The free fraction of (S)-propranolol in plasma was about four times larger than that of (R)-propranolol, and this is the main factor responsible for the differences in kinetics between the two enantiomers. There was a suggestion for a difference in tissue binding between the two enantiomers. With aging, the plasma and tissue concentrations of both enantiomers increase, probably because of a decrease in blood clearance. Tissue binding did not change much with aging. Notwithstanding the marked differences between the kinetics of the propranolol enantiomers, the changes that occur with aging affect both enantiomers to the same degree.

The plasma disposition of the enantiomers of ibuprofen has been investigated following oral administration of the racemic drug (400 mg) to 24 healthy male volunteers.[25] The plasma elimination of (R)-ibuprofen was found to be more rapid than that of the (S)-enantiomer (plasma half-life: (R) 2.03 hours; (S) 3.05 hours; 2P<0.001), resulting in a progressive enrichment in the plasma content of this isomer, some 64% of the total AUC being due to the pharmacologically active enantiomer. The influence of dose on the pharmacokinetic characteristics of the enantiomers of ibuprofen, over the range of 200 to 800 mg, was investigated in three subjects. Examination of dose-normalized AUC values and oral clearance indicates the dose dependence of (R)-ibuprofen disposition.

The enantioselective protein binding of mephobarbital (MPB) was investigated in human plasma and human serum albumin solutions by equilibrium dialysis.[26] A small but statistically significant difference was observed in the in vitro plasma protein binding of the enantiomers; (S)-MPB was ~59% bound and (R)-MPB 67% bound. The binding to albumin [(S)-MPB: 29% bound and (R)-MPB ~41% bound] was less than to plasma proteins but showed somewhat greater enantioselectivity, suggesting that albumin binding is a major source of the enantioselectivity in plasma. The effects of MPB concentration, of varying enantiomeric concentration ratio, and of phenobarbital on the enantioselective binding of MPB were studied. The effect of age was also investigated by measuring the binding in plasma from eight young (18- to 25-year-old) and eight elderly (60-plus-year-old) male subjects who took single doses of MPB. The results were in close agreement with the in vitro binding data, and the binding of both enantiomers was marginally but significantly lower in the young compared with the elderly subjects. These differences in binding were consistent with previously observed pharmacokinetic differences between the two subject groups.

Mefloquine (MQ) is a chiral antimalarial agent effective against chloroquine-resistant *Plasmodium falciparum*.[27] It is commercially available as a racemic mixture of the (+) and (−) enantiomers for oral administration. The pharmacokinetics of the (+) and (−) enantiomers of MQ were studied in eight healthy volunteers after administration of a first oral dose of 250 mg of racemic MQ and at steady state after 13 repeated doses of 250 mg given at 1-week intervals. Plasma samples were collected, and concentrations of each enantiomer were determined using a previously described achiral-chiral double column-switching liquid chromatographic method. At each time point, higher plasma concentrations values were found for the (−) enantiomer ($P<.001$). At steady state, C_{max} values of (−)-MQ were higher than those of (+)-MQ (1.42 ± 0.19 versus 0.26 ± 0.05 mg/L; $P<.001$). Similarly, the plasma concentrations 7 days after the final dose were higher for (−)-MQ (1.01 ± 0.26 versus 0.11 ± 0.04 mg/L; $P<.001$). AUC values at steady state were also higher for (−)-MQ (197.3 ± 36.7 versus 30.1 ± 8.9 mg/L × h; $P<.001$). The terminal half-life values ($t_{1/2b}$) were longer for (−)-MQ (430.4 ± 225.2 versus 172.8 ± 56.5 h; $P<.001$). This study shows that the pharmacokinetics of MQ is highly stereoselective.

A promising uricosuric, diuretic, and antihypertensive agent, 5-dimethyl-sulfamoyl-6,7-dichloro-2,3-dihydrobenzofuran-2-carboxylic acid (DBCA), was administered intravenously to rats.[28] The levels of DBCA in plasma and the AUC values of the S(−)-enantiomer were higher than those of the R(+)-enantiomer. Total body clearance was significantly greater for the R(+)-enantiomer. This stereoselective elimination was due to a difference in the nonrenal clearance, which seemed to reflect hepatic metabolism or biliary excretion. Hepatic metabolism seemed more likely because AUC and the amount of urinary excretion of the N-monodemethylated metabolite of DBCA were greater for the R(+)-enantiomer. The plasma had higher free fractions of the S(−)-enantiomer, a result suggesting that this enantiomer is distributed more readily to the tissues, including the liver. This result indicates that protein binding was not responsible for the stereoselective metabolism of (R)-(+)-DBCA. Although there was no difference in the renal clearances of the enantiomers, the renal clearance of free (R)-(+)-DBCA exceeded that of the S(−)-enantiomer, a result indicating the preferential excretion of the R(+)-enantiomer into the urine. Comparison of the pharmacokinetics of individual enantiomers after intravenous administration of each enantiomer or its racemate showed that the enantiomers interact with one another; dosing with the racemate delayed the elimination of each enantiomer because of mutual inhibition of hepatic metabolism and renal excretion for (R)-(+)-DBCA and of renal excretion for (S)-(−)-DBCA.

Detoxification

The detoxification of the enantiomers of glycidyl 4-nitrophenyl ether (GNPE), (−)-(R)- and (+)-(S)-GNPE, and glycidyl 1-naphthyl ether (GNE), (−)-(R)- and (+)-(S)-GNE, by rat liver glutathione transferase and epoxide hydrolase has been studied.[29] Enantioselectivity is observed with both enzymes favoring the (R)-isomers as determined by the formation of conjugate, diol, and the remaining substrate measured by HPLC. Enantiomers of GNE were detoxified by cytosolic epoxide hydrolase, but those of GNPE were not. Substantial nonenzymatically formed conjugates of enantiomers of GNPE were detected, showing (S)-GNPE the more reactive of the pair.

The influence of a single oral dose of 30 mg nicardipine on the pharmacokinetics of (R)- and (S)-propranolol, given orally as rac-propranolol 80 mg, was studied in 12 healthy volunteers.[30] The plasma concentrations were higher for the (S)-enantiomer than for the (R)-enantiomer. The Cl_o and the Cl'_{intr} of (S)-propranolol were significantly lower than the Cl_o and Cl'_{intr} of (R)-propranolol. The unbound fraction of (R)-propranolol was significantly higher than that of (S)-propranolol. These changes were more important for (R)- than for (S)-propranolol. The protein binding was not altered by nicardipine. The enantioselective effect of nicardipine on the metabolic clearance of propranolol appears to be due to an interaction at the level of the metabolizing enzymes. The effect of racemic propranolol on blood pressure was little affected when nicardipine was coadministered, and its bradycardic effect was reduced.

A surprisingly large number of marketed drugs are racemic mixtures. The pharmacokinetic literature on racemic drugs contains a vast amount of information on drug interactions derived from the measurement of total drug concentrations in plasma and urine. The appreciation of the role of stereochemistry in drug interactions with racemic warfarin resulted in a long-overdue scientific rigor being applied to the study of drug interactions. It is compelling that much of the literature was uninterpretable. A better understanding of oxidative metabolism, particularly the complexity of the cytochrome P-450 family of enzymes, has also strengthened the scientific basis of drug interactions. In an effort to more effectively use new and potent drugs, investigators and clinicians must consider both stereoselectivity and isozyme selectivity in the study of drug interactions, to understand the nature of the interaction.[31]

A solvent mixture containing dioxane, acetonitrile, and hexane was found to be suitable as a mobile phase to resolve oxazepam enantiomers by chiral stationary phase high performance liquid chromatography using covalent Pirkle columns.[32] The resolved oxazepam enantiomers (OX enantiomers) in this solvent mixture had a racemization half-life greater than 3 days at 23°C. When desiccated at 0°C as dried residue, OX enantiomers were stable for at least 50 days with less than 2% racemization. The conditions that stabilized OX enantiomers significantly facilitated the determination of racemization half-lives of OX enantiomers in a variety of aqueous and nonaqueous solvents and at different temperatures.

Chlorthalidone (CTD), a diuretic and antihypertensive agent, is a 3-hydroxy-3-phenylphthalimidine (HPP) carrying substituents on the phenyl group.[33] Chromatographic investigation of the degradation pathways of CTD showed the formation of a typical set of products. With one exception, the resulting chromatographic peaks were traced to previously identified compounds; the unknown component was found to be in equilibrium with CTD in aqueous media and has been identified as a dehydrated form, D^2-CTD.

The presence of the minor equilibrium product (ca. 0.7% at room temperature) spurred its investigation as a potential intermediate in the racemization of CTD. On the basis of structure alone, the facile racemization of the enantiomers of CTD was postulated, and it was corroborated by a recent report of CTD racemization catalyzed by acid or base. The study focused on the kinetics and activation parameters of a tripartite equilibrium found to exist between the enantiomers of CTD and achiral D^2-CTD through a planar carbonium ion intermediate; all reactions proceed under conditions approximating pure water at room temperature.

The enantiomers of chlorthalidone and a minor achiral dehydration product, D^2-CTD, have been shown to exist in dynamic equilibrium in aqueous media through a carbonium ion intermediate. The barrier to inversion at carbon is low for an uncatalyzed system: $DG = 21.6$ kcal/mol. This behavior extends to other 3-hydroxy-3-phenylphthalimidines (HPPs), with formation of the analogous D^2-HPP blocked by alkyl substitution at the 2-nitrogen.

The presence of one or several elements of chirality (centers, axes, or planes of chirality, and generally helicity) in drug molecules generates specific properties that may be advantageous in some cases, but inevitably require special attention. Examples of advantages include the possibility of increased selectivity and the fact that chirality, per se, is an invaluable probe in molecular pharmacology and biochemistry. In contrast, problems generated by stereoisomerism include the need for stereospecific analytical methods, the influence of the degree of resolution on activity, and the increased complexity of metabolic, pharmacologic, and clinical studies.

One major problem arising from the presence of elements of chirality is their possible lack of configurational stability—in other words, the danger of interconversion of stereoisomeric drugs. At a symposium titled "Chirality at the Crossroads" in Paris on April 26–29, 1992, the problem of the interconversion of stereoisomers was often mentioned and discussed.[12] During the symposium, it rapidly became obvious that the use of this term suffers from semantic confusion and even poor understanding, resulting in unproductive discussions and suggesting the possibility of inappropriate regulations. Clearly, configurational instability requires clarification. By classifying and discussing recognized examples, a few predictive rules can be offered, which should be an incentive for much-needed systematic investigations.

All cases considered involve the configurational instability of a single element of chirality either in compounds that contain only one such element (i.e., interconversion between enantiomers, chiral inversion) or in compounds that contain two or more such elements (i.e., interconversion between epimers, epimerization). This is the first criterion of classification used here. The second criterion separates nonenzymatic from enzymatic reactions. In addition, a distinction is made based on the nature of the unstable element of chirality.

A number of stereogenic elements in molecules may not be configurationally stable. Thus, while as a rule, tetracoordinate chiral centers are stable, tricoordinate chiral centers of first-row atoms (C, O, and N) are not. Chiral axes and chiral planes cover a continuum of possibilities between high lability and high stability. Attention needs to be focused on (a) tetracoordinate chiral carbon atoms displaying decreased stability and (b) helicity resulting from a nonsymmetrical, nonplanar ring system.

A variety of reactions can be categorized under the global concept of interconversion of stereoisomers. Thus, racemization or epimerization can result from inversion of labile chiral centers. From the examples available, some predictive rules are suggested for a chiral center of the type R″R′RC-H undergoing base-catalyzed inversion and provisional examples of affecting groups have been developed. Unimolecular inversion of nonsymmetrical, nonplanar ring systems can also result in racemization or epimerization, but no generalization can yet be offered. Besides these cases of nonenzymatic reactions, a limited variety of enzymatic reactions can operate to interconvert stereoisomers, the outcome rarely being a racemic mixture. An important aspect of stereoisomer interconversion is the time scale in which the phenomenon is observed. Thus, several reactions to nonenzymatic racemization or epimerization are fast when compared with the duration of action of the drug and therefore have pharmacologic significance, while others are slower and are of pharmaceutical relevance only.

It must be remembered that configurational stability and instability are relative phenomena. Given proper conditions (temperature and pH), no stereoisomer is configurationally stable. However, only two time scales and related sets of conditions are relevant as far as drugs are concerned. The pharmaceutical time scale and conditions imply that drugs remain (configurationally) stable during the whole manufacturing process and the shelf life, whereas the pharmacologic time scale is concerned with stability under physiological conditions (37°C, pH 7.4) and for the time the drug is in the body. Emphasis is generally on the pharmacologic time scale, but most of the information available is found in reports of pharmaceutical investigations.

Stereoisomeric enrichment occurring in vivo, particularly in enrichment of a stereoisomeric impurity, should not be confused with interconversion of stereoisomers. Indeed, differences in rates of metabolism or excretion may profoundly affect the stereoisomeric ratio of the drug in blood or in urine. Assessment of mass balance, or the use of compounds of very high stereoisomeric purity, should allow such confusions to be avoided.

To bring the problem of racemization into an even broader context, it must be recalled that the stereoisomeric purity of some amino acid residues in a number of proteins is known to be decreased in older people. Thus, both collagen and proteoglycan show an age-dependent accumulation of D-aspartic acid. This problem of "protein racemization" as it is (not quite properly) labeled is now beginning to attract attention as a possible mechanism or marker of tissue aging.

Stability

An example of a drug displaying extreme chiral instability is oxazepam, a drug whose kinetics of racemization were studied in organic solvents and aqueous solutions.[12] These studies are interesting and relevant from a pharmacologic viewpoint, since they indicate that oxazepam racemizes at ambient temperature and in the neutral pH range with a pseudo-first-order rate constant of 0.1 ± 0.05 min.$^{-1}$ (a half-life of racemization of approximately 10 ± 5 min.), suggesting a half-life of racemization of 1 to 4 min. at 37°C. This rate of racemization is extremely fast compared to the duration of action of the drug, indicating that oxazepam is correctly viewed as a single compound existing in two very rapidly interconverting chiral states.

Kinetic information is also available for the anorectic agent amfepramone (also known as diethylpropion) and its two N-deethylated metabolites, N-ethylaminopropiophenone and aminopropiophenone (also known as rac-cathinone). Assuming a common intermediate for racemization and α-carbon deuteration, the rate of the latter reaction was measured in D_2O (35°C and pH 7.3–7.4), revealing $t_{1/2}$ values of 15 and 21 hours for the tertiary and secondary amine, respectively. The half-life of racemization of the primary amine was estimated to be about 30 hours. These values suggest that configurational inversion should have only a marginal influence on the pharmacologic activity of these compounds, their biological half-lives being considerably shorter.

The extent of racemization of (+)-chlorthalidone as a function of pH has been examined. The minimum of the log K/pH curve is pH 3.[34] The reaction mechanism of inversion is postulated to involve a carbenium cation over the entire pH range and a ring opening reaction in the alkaline range.

Because demand for optically pure drugs is continually increasing, it is of great importance to elucidate factors affecting stereochemistry in order to provide a stable formulation with a high chiral quality of the desired isomer.[34] Isomerization may be influenced by factors such as pH, buffer salts, ionic strength, solvents, temperature, and light. One of the aims of stability studies of optical isomers in drug formulations is therefore to determine the extent of isomerization. Suitable and specific methods must be employed, and limitation of the undesired isomers should be specified.

Because of stereoselective interactions with receptors, we know that enantiomers can exhibit different biological actions and that the distomer can even give rise to adverse effects, as the example of penicillamine demonstrates.

Guidelines have now been issued by the EEC and FDA, which include rules concerning, for example, batch-to-batch consistency of the enantiomeric ratio (enantiomer purity) and a full description of the stability of the drug substance (enantiomeric stability). These requirements are also applicable to pure stereoisomeric drug substances in pharmaceutical dosage forms, since optical isomers are not invariably stable compounds.

Only a few systematic studies have been performed to elucidate the stability of the configuration of chiral drugs in pharmaceutical dosage forms. For example, Lamparter et al.[34] have thoroughly investigated the influence of α-cyclodextrins and β-cyclodextrin derivatives on the racemization behavior of tropane alkaloids.

Industrial Perspectives

The current position of the Pharmaceutical Manufacturers Association is that the development of either a racemate or single enantiomer should be made on a case-by-case basis, depending on pharmacologic and toxicologic considerations and technical feasibility, and that clinical and preclinical data obtained on racemates may be used to support the development and marketing of a single isomer.[35]

A sponsor's decision to develop, register, and market a racemate must be scientifically justified. The potential risk and benefit to the patient will weigh significantly in this decision-making process. Although not all-inclusive, the following specific examples illustrate typical situations where the decision to develop a racemate might be made:

- The enantiomers have been shown to have pharmacologic and toxicologic profiles similar to the racemate.
- The enantiomers are rapidly interconverted in vitro or in vivo on a time scale such that administering a single enantiomer offers no advantage.
- One enantiomer of the racemate is shown to be pharmacologically inactive and the racemate is demonstrated to be safe and effective.
- Synthesis or isolation of the preferred enantiomer is not practical. This assumes that meaningful effort has been directed to synthesis and isolation of the preferred enantiomer without success or that even though isolation and synthesis may have been realized on a small scale, large-scale application of these methods may not be feasible.
- Individual enantiomers exhibit different pharmacologic profiles, and the racemate produces a superior therapeutic effect relative to either enantiomer alone.

If a drug product contains a chiral drug substance, regardless of its stereochemical form (racemate or enantiomer), and has not previously undergone registration in the same or in another stereochemical form, it should be regarded as a new chemical entity (NCE).[36] The registration of any NCE requires comprehensive preclinical and clinical study; the details of the study are generally established on a case-by-case basis.

The choice of the stereochemical form of a chiral drug should be based on scientific data relating to quality, safety, efficacy, and risk/benefit. The responsibility for the decision as to which stereochemical form of a drug might be developed should clearly reside with the applicant. Regulatory requirements concerning the development of chiral drugs should be consistent with these principles.

After a drug has received approval to be marketed in a particular stereochemical form, subsequent comparative clinical study of individual enantiomers should not be required automatically unless a new safety issue that may be dependent on stereochemical configuration arises or new claims are envisaged that are to be based on the pharmacologic or toxicologic activities of one or both of the enantiomers. An exception to this might be an agreement by the applicant to conduct such studies as a condition to obtaining marketing authorization.

When a stereochemical form of a chiral drug is chosen for development, a validated enantiospecific assay for use in dosage-form development and an assay of enantiomers in biological fluid should be developed at an early stage and used unless a nonenantioselective assay(s) provides results equivalent to those obtained with the enantiospecific assay. The pharmacokinetic assays used in the study of chiral drugs should be capable of accurately measuring levels of individual enantiomers in biological fluids, unless it has been shown that stereoselectivity effects are not significant in the pharmacokinetics of the drug.

For a single enantiomer, susceptibility to stereochemical instability via either in vivo or in vitro inversion, racemization, or epimerization should be established. For a racemate, the potential for in vivo stereochemical inversion of one of the enantiomers must be assessed. The in vitro stereochemical stability of the drug substance should be evaluated throughout the claimed shelf life of the product.

The assays and purity tests that are applied to the bulk drug substance and the dosage form and that are used in release and stability testing should be capable of accurately measuring both the content of the drug substance and the content of all relevant stereochemically related and chemically related substances unless the absence of stereoselective degradation has been demonstrated.

Purity evaluation of a drug containing a single enantiomer, the optical antipode, should be treated as a potential related substance, impurity, or degradation product and addressed as such in accordance with normal practices. The purity level required for a chiral reference standard should depend on its intended use. An absolute measure of enantiopurity should be stated when a reference standard will be used as an assay standard.

It is important to demonstrate the batch-to-batch consistency by evaluating the impurity profile of a chiral drug substance, especially with respect to the content of chemically and stereochemically related substances in batches used for toxicology studies and clinical trials and their relationship to those in the marketed drug product.

Enantiospecific identity tests preferably should be used for the identification of a chiral drug, regardless of its stereochemical form. A validated optical rotation measurement is desirable unless other methods have been found to be more suitable. When describing the production process for a chiral drug substance, the synthetic steps in which the stereogenic center is generated, maintained, affected, or manipulated should be provided in relevant detail in the marketing application.

During development of a formulation form of a chiral drug, regardless of its stereochemical form, it is important to consider the effect of its stereochemical form on the physical-chemical properties of the drug substance (polymorphism, rate of dissolution, crystallinity). For a racemate, it may be useful to determine if the drug substance is a racemic compound (i.e., a true racemate) or a racemic mixture (a conglomerate). Generic drug products that contain a chiral drug substance should be held to the same manufacturing and control standards that were originally applied in the approval of the innovator's product. Discussed next is development of a racemate and racemate enantiomer switch.[36]

Development of Racemates

Let's consider the case of a racemate that, despite being composed of two optical antipodes, need not be differentiated from other NCEs containing a single chemical species. While production of a single enantiomer may be achieved on a small scale, scale-up may not always be possible; and if scale-up problems contribute to the decision to develop a racemate, a discussion of the work undertaken to attempt scale-up should be provided in the marketing application. Otherwise, the justification for the development of a racemate should generally not be required.

The pharmacodynamic profile of each enantiomer should be established for primary pharmacologic effect with establishment of secondary effects as required to assure safety. If a significant or unexpected toxicity is observed in preclinical studies with a racemate, its relationship to one or both of the enantiomers must be investigated.

Animal safety studies should be supported by toxicokinetics that employ an enantiospecific assay in order to assess enantiomer levels to assure that the animals' exposure to each enantiomer supports the anticipated exposures in human studies. If the pharmacokinetic profile of the enantiomers is the same or a fixed ratio in various animal species, the use of a nonenantiospecific assay may be appropriate for subsequent studies.

The clinical pharmacokinetics of both enantiomers should be defined at an early stage, after administration of the racemate, and these data should be evaluated in the context of results of preclinical toxicokinetic studies.

The use of nonenantiospecific assays in bioequivalence studies may be justified only when studying single stable enantiomers or, in the case of a racemate, when it has been shown that the rate of release of the enantiomers from the formulation does not influence the relative AUC values.

Racemate-Enantiomer Switches

A drug product that contains a chiral drug substance, regardless of its stereochemical form, that has been approved in another stereochemical form may represent a special case requiring less comprehensive preclinical and clinical development than would ordinarily be required for an NCE. The details of such an abbreviated development program must be established on a case-by-case basis and justified by the applicant. For example, for a racemate-to-enantiomer switch, it may not be necessary to repeat certain studies that would normally be required with the selected enantiomer when relevant studies have previously been conducted with the racemate and the data from these studies are available to the applicant of the enantiomer. Additional studies of relevant stereochemical species (bridging studies), which are determined on a case-by-case basis, may be required to support the elimination of certain normally required studies. The feasibility of conducting such studies will depend on the stereochemical integrity of the enantiomers to be investigated. Examples of these bridging studies are:

1. Comparisons of the profiles of significant pharmacologic and toxicologic activities of the selected enantiomer and the racemate (it may also be necessary to study the optical antipode in order to investigate the possibility of interactions)
2. Comparisons of the pharmacokinetic profiles of the selected enantiomer and the racemate (it may also be necessary to study the pharmacokinetics of the optical antipode in order to assure that its removal does not alter the kinetics of the single enantiomer)

An acceptable preclinical toxicology bridging strategy might consist of a three-month repeat dose comparison of the selected enantiomer and the racemate in the most appropriate species and a segment II reproductive toxicity comparison in the most appropriate species. Where a single enantiomer shows evidence of increased toxicity compared to that of the racemate, this should be further investigated and the implications of this toxicity for therapeutic use must be considered.

Since the effects of the addition of the optical antipode on the overall characteristics of a single enantiomer-based drug product would be unknown, the switch of an approved single enantiomer-based drug to a racemate-based drug should be approached as if it were the development of an unrelated NCE.

Regulatory agencies must provide clear regulatory guidance to promote the efficient development of new drugs. However, science should lead the development of regulations and not vice versa. As the result of recent activity concerning drug stereochemistry on the part of regulatory agencies around the world and the global approach to drug development that many transnational companies now pursue, there exist a need and an opportunity to achieve a sensible consensus in the establishment of regulatory requirements for chiral drugs.

Regulatory Guidelines

In 1987, the FDA issued a set of initial guidelines on the submission of new drug applications (NDAs), where the question of stereochemistry was approached directly in the guidelines on the manufacture of drug substances.[37, 38] These guidelines were finally released in 1992.[39] They require a full description of the methods used in the manufacture of the drug, including testing to demonstrate its identity, strength, quality, and purity (discussed below). In short, the submissions should show the applicant's knowledge of the molecular structure of the drug substance. For chiral compounds, this includes identification of all chiral centers. The enantiomer ratio, although 50:50 by definition for a racemate, should be defined for any other admixture of stereoisomers. The proof of structure should consider stereochemistry and provide appropriate descriptions of the molecular structure. An enantiomeric form can be considered an impurity, and therefore it is desirable to explore potential in vivo differences between these forms.

Some guidance on regulatory practices has been provided in Canada, EEC, Japan, and the United States in documents at various stages of development. All of the requirements are based on similar scientific bases. These requirements evolved after discussions among various authorities with industry and among international colleagues. As a result, the texts are comparable to each other in content.

The current regulatory position of the FDA with regard to the approval of racemates and pure stereoisomers are discussed under the following situations:[37-40]

1. Circumstances in which stereochemically sensitive analytical methods are necessary to ensure the safety and efficacy of a drug have been discussed
2. Regulatory guidelines for NDAs are interpreted for the approval of a pure enantiomer in which the racemate is marketed, for the approval of either a racemate or a pure enantiomer in which neither is marketed, and for clinical investigations to compare the safety and efficacy of a racemate and its enantiomers

The basis for such regulations has been drawn from historical situations (thalidomide and benoxaprofen) as well as currently marketed drugs (arylpropionic acids, disopyramide, and indacrinone). The primary regulatory focus of the FDA is on considerations of both clinical efficacy and consumer safety of a potential drug. Because the chiral environment found in vivo affects the biological activity of a drug, the approval of stereoisomeric drugs for marketing can present special challenges. As mentioned before, the case of thalidomide is an example of a problem that may have been complicated by ignorance of stereochemical effects. The use of racemates can lead to erroneous models of pharmacokinetic behavior and to the potential for opportunities to manipulate pharmacologic activity. It is technically feasible to design experiments that will unambiguously determine whether a stereochemically pure drug is more effective and/or less toxic than the racemate.

The federal Food, Drug and Cosmetic Act requires a full description of the methods used in the manufacture of the drug, which includes testing to demonstrate its identity, strength, quality, and purity. The question of stereochemistry was approached directly in the guidelines issued by the FDA in 1987 on the submission of NDA for the manufacture of drug substances.[40] Therefore, the submissions should show the applicant's knowledge of the molecular structure of the drug substance. The guidelines do not discuss conditions under which a determination of absolute configuration is desirable or essential. Obviously, it would be appropriate information for supporting the manufacture of optically pure drugs.

U.S. regulatory requirements demand that the bioavailability of the drug be demonstrated.[41] When pharmacokinetic models differ between enantiomers, it seems obvious that establishing the bioavailability of the drug from a racemate is a much more complex task, which cannot be accomplished without separation of the enantiomers and investigation of their pharmacokinetics as individual molecular entities.

The toxicity of impurities, degradation products, and residues from manufacturing processes should be investigated as a drug is being developed. The same standards should, therefore, be applied to the enantiomeric molecules in a racemate.[39] Whenever a drug can be obtained in a variety of chemically equivalent forms (such as enantiomers), it makes sense to explore the potential in vivo differences between these forms.

Although it is now technologically feasible to prepare purified enantiomers, development of racemates may continue to be appropriate. However, the following should be considered in product development:

1. Appropriate manufacturing/control procedures should be used to assure stereoisomeric composition of a product with respect to identity, strength, quality, and purity. Manufacturers should notify compendia of these specifications and tests.
2. Pharmacokinetic evaluations that do not use a chiral assay will be misleading if the disposition of the enantiomers is different. Therefore, techniques to quantify individual stereoisomers in pharmacokinetic samples should be available early. If the pharmacokinetics of the enantiomers are demonstrated to be the same or to exist as a fixed ratio in the target population, an achiral assay or an assay that monitors one of the enantiomers may be used subsequently.

Selective methodology must be developed for resolving and monitoring stereoisomers, and determining their absolute configuration would be useful. Method development is described in great detail in chapter 8; examples of determination of absolute configuration are given in chapter 3.

REFERENCES

1. First International Symposium on Separation of Chiral Molecules, Paris, May 31–June 2, 1988.
2. S. Ahuja, *Chiral Separations: Applications and Technology.* American Chemical Society: Washington, DC, 1997.
3. H. Y. Aboul-Enein and I. W. Wainer, eds., *The Impact of Stereochemistry on Drug Development and Use.* Wiley: New York, 1997.
4. S. Ahuja, *Chiral Separations by Liquid Chromatography*, ACS Symposium Series #471. American Chemical Society: Washington, DC, 1991.
5. S. Allenmark, *Chromatographic Enantioseparation.* Ellis Horwood: New York, 1991.
6. S. Ahuja, *Selectivity and Detectability Optimizations in HPLC.* Wiley: New York, 1989.
7. W. J. Lough, *Chiral Liquid Chromatography.* Blackie and Son: Glasgow, UK, 1989.
8. J. Gal, *LC-GC*, **5,** 106 (1987).
9. S. Hara and J. Cazes, *J. Liq. Chromatogr.*, **9,** Nos. 2 and 3 (1986).

10. R. W. Souter, *Chromatographic Separations of Stereoisomers.* CRC Press: Boca Raton, FL, 1985.
11. G. Blaschke, H. P. Kraft, K. Fickentscher, and F. Koehler, *Arzneim.-Forsch*, 29, 1640 (1979).
12. B. Testa, P. A. Carrupt, and J. Gal, *Chirality*, 5, 105 (1993).
13. "Top 200 Drugs in 1982," *Pharmacy Times*, p. 25 (April 1982).
14. Y. Okamoto, *CHEMTECH*, 176 (March 1987).
15. S. Ahuja, Eastern Analytical Symposium, Somerset, NJ, November 17–22, 1996.
16. J. S. Millership and A. Fitzpatrick, *Chirality*, 5, 573 (1993).
17. A. J. Hutt, *Chirality*, 3, 161 (1991).
18. B. Waldeck, *Chirality*, 5, 355 (1993).
19. B. B. Barlow, F. M. Franks, and J. D. M. Pearson, *J. Pharm. Pharmacol.*, 24, 753 (1972).
20. M. R. Wright, N. M. Davies, and F. Jamali, *J. Pharm. Sci.*, 83, 911 (1994).
21. A. B. Malmberg and T. L. Yaksh, *Science*, 257, 1276 (1992).
22. D. R. Brocks, W. T. C. Liang, and F. Jamali, *Chirality*, 5, 61 (1993).
23. S. M. Sannins, W. J. Adams, D. G. Kaiser, G. W. Halstead, J. Hosley, H. Barnes, and T. A Baillie, *Drug Metab. and Disposition*, 19, 405 (1991).
24. A. M. Vermulen, F. M. Belpaire, E. Moerman, F. DeSmet, and M. G. Bogaert, *Chirality*, 4, 73 (1992).
25. A. Avgerinos and A. J. Hutt, *Chirality*, 2, 249 (1990).
26. N. J. O'Shea and W. D. Hooper, *Chirality*, 2, 257 (1990).
27. F. Gimenez, R. A. Pennie, G. Koren, C. Crevoisier, I. G. Wainer, and R. Farinotti, *J. Pharm. Sci.*, 83, 824 (1994).
28. K. Higaki, K. Kadono, and M. Nakano, *J. Pharm. Sci.*, 81, 935 (1992).
29. R. Chen, P. Nguyen, Z. You, and J. E. Sinsheimer, *Chirality*, 5, 501 (1993).
30. I. Vercruysse, F. Belpaire, P. Wynant, D. L. Massart, and A. G. Dupont, *Chirality*, 6, 5 (1994).
31. M. Gibaldi, *Chirality*, 5, 407 (1993).
32. S. K. Yang and X. L. Lu, *Chirality*, 4, 446 (1992).
33. G. Severin, *Chirality*, 4, 226 (1992).
34. E. Lamparter, G. Blaschke, and J. Schluter, *Chirality*, 5, 370 (1993).
35. PMA Ad Hoc Committee on Racemic Mixtures, *Pharm. Technology*, 46 (May 1990).
36. M. Gross, A. Cartwright, B. Campbell, R. Bolton, K. Holmes, K. Kirkland, T. Salmonson, and J-L. Robert, *Drug Information Journal*, 27, 453 (1993).
37. A. G. Rauws and K. Gruen, *Chirality*, 6, 72 (1994)
38. W. H. DeCamp, *Chirality*, 1, 2 (1989).
39. FDA's Policy Statement for the Development of New Stereoisomeric Drugs, May 1, 1992; Corrections made on January 3, 1997. http://www.fda.gov/cder/guidance/stereo.htm
40. *Guidelines for Submitting Supporting Documentation in Drug Applications for the Manufacture of Drug Substances.* Office of Drug Evaluation and Research (HFD-100), Food and Drug Administration: Rockville, MD, 1987.
41. Code of Federal Regulation, Title 21, Government Printing Office, Sec. 314.5(d)3: Washington, DC, 1988.

REVIEW

1. Discuss why it is necessary to study pharmacology and toxicology of each enantiomer used in humans.

2. Explain why it is not always necessary to market single enantiomers

TUTORIAL

A clear understanding of the FDA's requirements on chiral compounds targeted as drug candidates is essential. Review the FDA's Policy Statement for the Development of New Stereoisomeric Drugs, May 1, 1992; Corrections made on January 3, 1997. http://www.fda.gov/cder/guidance/stereo.htm. FDA's perspective is summarized below.

- All chiral centers should be identified.
- The enantiomeric ratio should be defined for any admixture other than 50:50.
- Proof of structure should consider stereochemistry.
- Enantiomers may be considered impurities.
- Absolute configuration is necessary for an optically pure drug.
- Marketing an optical isomer requires an NDA.
- IND is required for clinical testing.
- Justification of the racemate or any of the optically active forms must be made with the appropriate data.
- Pharmacokinetic behavior of the enantiomers should be investigated.

3

Review of Stereochemistry

Stereochemistry (a word derived from the Greek prefix *stereos*-, meaning solid) can be best defined as the study of chemical compounds in three dimensions. Since most molecules are three-dimensional, stereochemistry is extremely important in helping us understand their structural interactions. Another important point to remember is that three-dimensional figures can exist as nonsuperimposable mirror images. For example, a mirror image of a right hand cannot be placed exactly on the top of a right hand, with both palms up or down. This leads to the statement that a chiral (from the Greek word *cheir*, meaning hand) object is not superimposable on its mirror image. Examples of nonsuperimposable objects shown in Figure 3.1 include a hand and a key.

An achiral object, on the other hand, is superimposable on its mirror image. A key without grooves or with identical grooves on both sides is achiral because it has a plane of symmetry. Other examples include a nail and a T-shirt without a pocket (Figure 3.2). A similar spatial relationship exists for molecules because molecules in nature exist as three-dimensional symmetrical or asymmetrical figures. The two forms of a chiral molecule with one asymmetric center are called enantiomers. Unlike other stereoisomers (see Stereoisomerism, later in this chapter), enantiomers have identical physical and chemical properties and are consequently difficult to separate and quantify.

This chapter provides a brief review of stereochemistry to enable readers to better understand the separation methods discussed in this book. For a detailed discussion of stereochemistry, you may also want to refer to some of the interesting books on the subject.[1-4]

Figure 3.1 Nonsuperimposable objects. Reproduced with permission from S. Ahuja, W. Pirkle, and C. Welch, *Chiral Separations by Chromatography*, American Chemical Society Short Course, 1994. Copyright 1994 American Chemical Society.

Brief History

A short historical picture is provided here to better acquaint the reader with the discovery of this fascinating field of chemistry that has achieved tremendous importance. In 1801, René-Just Haüy, a French mineralogist, noticed that quartz crystals are hemihedral.[5] The significance of this observation is that certain facets of the crystals are so disposed as to produce nonsuperimposable species, which are related as an object to its mirror image. These mirror-image crystals are called "enantiomorphous," from the Greek words *enantios* and *morphe*, meaning opposite form.

The origin of stereochemistry can be related to the discovery of plane-polarized light by the French physicist Étienne-Louis Malus[6] in 1809 (see Enantiopurity, later in this chapter). This led to the discovery of optical rotation by French scientist Jean-Baptiste Biot in 1812,[7] following an observation a year earlier by his colleague François Arago. Malus observed that a quartz plate cut at right angles proportional to its crystal axis rotates the plane of polarized light through an angle proportional to the thickness of the plate. In 1815, he extended these observations to organic substances in a liquid state, such as turpentine, and solutions of solids, such as sucrose or tartaric acid. He recognized the difference in rotation produced by quartz and that produced by the organic substances studied by him.

John Frederick Herschel, a British astronomer, observed in 1822 that there was a relation between hemihedrism and optical rotation:[8] All quartz crystals having the odd faces inclined in one direction rotate the plane of polarized light in one and the same sense, whereas enantiomorphous crystals rotate polarized light in the opposite sense.

Figure 3.2 Achiral objects. Reproduced with permission from S. Ahuja, W. Pirkle, and C. Welch, *Chiral Separations by Chromatography*, American Chemical Society Short Course, 1994. Copyright 1994 American Chemical Society.

In 1848, Louis Pasteur succeeded in separating crystals of sodium ammonium salts of (+)- and (−)-tartaric acid from the racemic mixture. Based on the analogy between crystals and molecules, Pasteur further recognized that the power to rotate polarized light was caused by dissymmetry—that is, by dissimilarity of the crystal or molecule and its mirror image, expressed in the case of the ammonium sodium tartrate crystals by the presence of hemihedric faces. Similarly, he postulated that the molecular structure of (+)- and (−)-tartaric acids must be related as an object to its mirror image.[9] The two acids can be considered as enantiomorphous at the molecular level and are generally called enantiomers. Here, the ending -mer, derived from the Greek word *meros*—meaning part and usually referring to a molecular species, is added to *enantio*.

Stereoisomerism

Stereochemical studies can be divided into two categories:

- Static stereochemistry focuses on molecules—their structure, energy, physical properties, and most of their spectral properties.
- Dynamic stereochemistry focuses on chemical reactions and their stereochemical requirements and outcomes, including interconversion of conformational isomers.

Most of our discussion in this chapter relates to static stereochemistry as it might be utilized in designing and performing better separations. Symmetry classifies stereoisomers as either enantiomers or diastereomers, which are optical isomers that are not related as an object and its mirror image. The most common diastereomeric molecule is one that contains two asymmetric carbons. Unlike enantiomers, the physical and chemical properties of diastereomers can differ significantly. It is not unusual for them to have different melting or boiling points, refractive indices, solubilities, and so on. Their optical rotation can differ in both sign and magnitude.

Other points to remember:[10]

- A pair of enantiomers is possible for all molecules containing a single chiral carbon atom (one with four different groups attached).
- Diastereoisomers, or diastereomers, are basically stereoisomers that are not enantiomers of each other.
- A chiral molecule can have only one enantiomer; however, it can have several diastereomers or none.
- Two stereoisomers cannot be both enantiomers and diastereomers of each other simultaneously.

Stereoisomerism can result from a variety of sources in addition to the single chiral carbon (stereogenic or chiral center)—a chiral atom that is a tetrahedral atom with four different substituents. The term *stereogenic center* is considered preferable to *chiral center*; however, both terms are used interchangeably in this book. It is not necessary for a molecule to have a chiral carbon in order to exist in enantiomeric forms, but it is necessary that the molecule, as a whole, be chiral.

There are two simple molecular sources of chirality: molecules with a stereogenic center and molecules with a stereogenic axis. Some discussion on this subject is provided in the section titled Configuration and Conformation. Detailed discussion on these topics may be found in several books and review articles;[1-4] a short summary is provided here.

Stereoisomerism is possible in molecules that have

- One or more centers of chirality
- Helicity (e.g., helical nature of tertiary structures of proteins, polysaccharides, and nucleic acids)
- Planar chirality (e.g., paracyclophanes)
- Axial chirality (e.g., spiranes with cyclic skeleton)
- Torsional chirality (e.g., torsion about double or single bonds like cis- and trans-isomers and rotamers)
- Topological asymmetry (e.g., catenanes)

Configuration and Conformation

Stereoisomers may be either configurational or conformational. *Configuration* relates to a particular spatial arrangement of the atoms in molecules of defined constitution without regard to arrangement, differing only in torsion around single bonds. *Conformation* deals with arrangement of the atoms in molecules of defined configuration resulting from torsion around one or more single bonds.

It is not possible to provide a clear concept of molecular structure or shape unless constitution, configuration, and conformation are defined.

Simple descriptors such as *cis* (when ligands are located on the same side of double bond or ring structure in conformation devoid of reentrant angles) and *trans* (opposite of *cis*) for geometric stereoisomers become ambiguous when applied to complex olefins, so a newer system based on the Cahn-Ingold-Prelog Convention has been used (see later in this chapter). The E,Z system examines the groups attached to one carbon of the double bond and arranges them in order of preference. The operation is repeated at the other carbon. The higher preference groups on two carbon are compared, and if they are on the same side of the double bond, the alkene is designated as Z (*zusammen*). If they are on the opposite side of the double bond, the alkene is designated as E (*entgegen*).

As mentioned earlier, the symmetry factor classifies stereoisomers as either enantiomers or diastereomers. The amounts of energy necessary to convert given stereoisomers into their isomeric forms may be used for their classification. Stereoisomers with low-energy barriers to this conversion are termed conformational isomers, whereas high-energy-barrier conversions are described as configurational isomers. Diastereomers differ in energy content, and thus in every physical and chemical property; however, the differences may be so minute as to be nearly indistinguishable.

Molecular Structure and Chirality

The discussion on this subject is covered from the standpoint of a single asymmetric atom or overall molecular structure.

Single asymmetric atom. Tetrahedral structures formed by four different groups around atoms of elements such as carbon, silicon, nitrogen, phosphorus, or sulfur are well-known examples of chiral compounds. Elements of groups V or VI in the periodic table can also form nonplanar structures with three ligands, where the lone-pair electrons of the central atom can be regarded as the fourth ligand. In this case, a planar configuration is always possible, the formation of which will permit interconversion between the chiral pyramidal structures. Some examples of chiral molecules are given in Figure 3.3, where the central atom is a carbon atom. Other examples are given in Figure 3.4, where the central atom is not a carbon atom.

Overall structure. Allene ($CH_2=C=CH_2$) provides a simple example of an overall chiral structure not involving an asymmetric central atom. Substitution of one of the hydrogen atoms at each of the carbon atoms by a substituent X is sufficient to generate chirality—that is, the geometry of the molecule will lead to a non-superimposable mirror image (Figure 3.5).

Metallocenes and metal ion alkene coordination complexes provide similar types of isomerism. The key element of this type of isomerism is molecular rigidity. Because no rotation around the double bond is possible under normal conditions, any interconversion between the optical isomers is eliminated. This discussion suggests that any structure of this type would generate stable optical isomers as long as interconversion through rotation were sufficiently restricted. This type of isomerism is called atropisomerism as defined next.

Atropisomerism. *Atropisomerism* refers to stereoisomerism resulting from restricted rotation about single bonds where the rotational barrier is high enough to

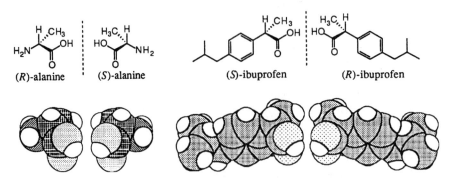

Figure 3.3 Chiral molecules with a central C atom. Reproduced with permission from S. Ahuja, W. Pirkle, and C. Welch, *Chiral Separations by Chromatography*, American Chemical Society Short Course, 1994. Copyright 1994 American Chemical Society.

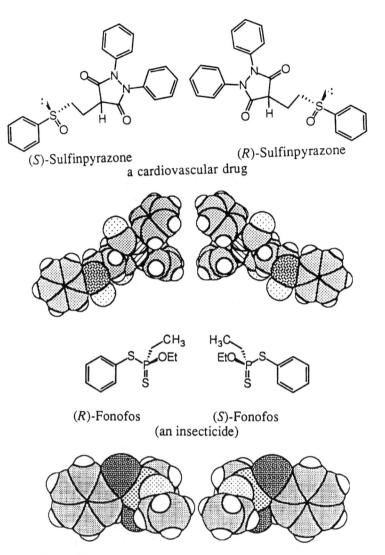

Figure 3.4 Chiral molecules where the central atom is not a C atom. Reproduced with permission from S. Ahuja, W. Pirkle, and C. Welch, *Chiral Separations by Chromatography*, American Chemical Society Short Course, 1994. Copyright 1994 American Chemical Society.

permit isolation of the isomeric species. Certain substituted biaryls — for example, o,o'-dinitrodiphenic acid — meet this requirement. Another example, β-binaphthol, is shown in Figure 3.6.

In the case of substituted biaryls, restricted rotation around the central carbon is caused by the steric effect of substituents. Configurational stability is provided because the groups are too large to allow rotation past each other.

Allenes

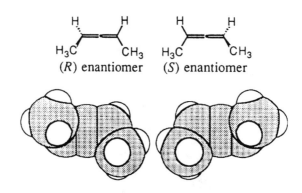

Extended Even Allenes

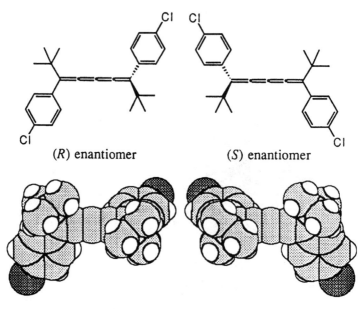

Figure 3.5 Substituted allenes. Reproduced with permission from S. Ahuja, W. Pirkle, and C. Welch, *Chiral Separations by Chromatography*, American Chemical Society Short Course, 1994. Copyright 1994 American Chemical Society.

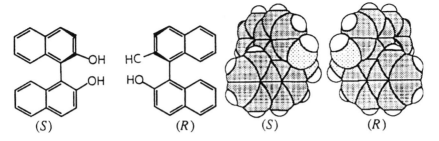

Figure 3.6 Atropisomers. Reproduced with permission from S. Ahuja, W. Pirkle, and C. Welch, *Chiral Separations by Chromatography*, American Chemical Society Short Course, 1994. Copyright 1994 American Chemical Society.

Twisted conformation. Polarized ethylenes provide an interesting example of "push-pull" effect, where electronic effects of substituents cause a weakening of the bond, thus lowering the barrier to rotation. The molecule will adopt a twisted conformation in solution if the steric interactions between the two halves of the molecules are strong enough. This can lead to chiral conformations of sufficient stability to permit resolution of enantiomers at room temperature. Compounds with partial double bond potential present a similar type of situation, as exemplified in Figure 3.7.

Steric crowding in a molecule may cause molecular distortion that could give rise to chirality. A very good example of this phenomenon is found in condensed aromatic ring hydrocarbons called helicenes, as shown in Figure 3.8.

Steric constraints can give rise to a helical form with an energy barrier to interconversion between the right- and left-handed enantiomers, which is high enough to permit their resolution.

Classification

We know that chiral molecules can have different natures and shapes; however, all of them can be classified into the following three categories:

- Central chirality: Three-dimensional space is occupied asymmetrically around a chiral center.
- Axial chirality: Three-dimensional space is occupied asymmetrically around a chiral axis.

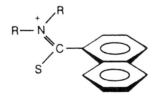

Figure 3.7 Twisted conformation.

Figure 3.8   Helicene.

- Planar chirality: Three-dimensional space is occupied asymmetrically around a chiral plane.

Nomenclature

A number of methods have been used over the years for naming chiral compounds. Some of these are still used or are encountered in literature. So it is beneficial to be familiar with them.

D-, L- Nomenclature

In this system, proposed by Emil Fischer[11] in 1891, bonds may be visualized as shown for D- and L-glyceraldehyde in Figure 3.9.

- Vertical bonds are directed backward from the plane of paper.
- Horizontal bonds are directed outward from the plane of paper.

In 1951, Bijovet et al. determined the absolute configuration of rubidium D-(+)-tartrate by X-ray crystallography, making use of an anomalous dispersion effect.[12] Coincidentally, the configuration of D-glucose, arbitrarily assigned by Fischer, is correct: it configurationally relates to D-glyceraldehyde.[13] This system of nomenclature, which relates a variety of optically active compounds to each other, especially carbohydrates, is still used. However, the system is applicable only to compounds having asymmetric carbon atoms and thus cannot be used as a general nomenclature system.

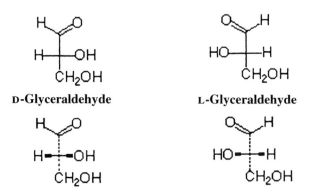

Figure 3.9   Fischer projections.

Cahn-Ingold-Prelog System

The basis of this system[14, 15] can be described as follows:

1. Arrange the ligands associated with the element of chirality in a sequence according to the following rules:
   a. Higher atomic number is given priority.
   b. Higher mass has priority.
   c. When the proximate atom of two or more of the ligands is the same, the atomic number of the next atom determines priority.
   d. Double or triple bonds are counted as if they were split into two or three bonds, respectively.
   e. *Cis* is given higher priority than *trans*.
   f. Like pairs [(R,R) or (S,S)] are given priority over unlike pairs [(R,S) or (S,R)].
   g. Lone-pair electrons are regarded as an atom with atomic number 0.
   h. Proximal groups take priority over distal groups.
2. Use this sequence to trace a chiral path. The ligands are arranged in a sequence C > D > E > F, and are then viewed in a way that F (lowest priority) is pointing backward from the viewer.
3. The remaining ligands are then counted, starting from the one of the highest priority. If the sequence is clockwise, then the designation is (R) for *rectus* (a word derived from French, meaning straight). If the sequence is counterclockwise, then it is designated (S) for *sinister* (a word derived from French, meaning left).

There are a number of disadvantages associated with the priority rules in the Cahn-Ingold-Prelog system. For example, in chiral epoxides, when the X substituent is changed from CH_3 to Cl, the absolute configuration changes from R to S (Figure 3.10).

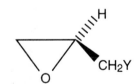

Y	Absolute configuration
CH_3	R
F	S
Cl	S
OH	R

Figure 3.10 Configuration of chiral epoxides.

M,P System

It is difficult to assign (R) or (S) configuration to molecular helixes. An alternative M,P system is often used (see Figure 3.11):

1. View the helix from one end.
2. Trace the path of the helix receding from the view.
3. A clockwise path denotes the plus (P) configuration.
4. A counterclockwise path denotes the minus (M) configuration.

Resolution

As discussed in chapters 1 and 2, it is frequently necessary to resolve enantiomers for a number of reasons. Resolution can be defined as separation of a racemate into its two enantiomer constituents. In the resolution methods, since the point of initiation is a racemate, the maximum potential yield of an enantiomer is 50%. Resolution methods may involve physical processes only or chemical reactions.

In resolutions mediated by chemical reactions, diastereomeric transition states or diastereomeric compounds are produced. To use these methods, it is necessary to take advantage of either thermodynamic or kinetic control. This necessitates development of appropriate methods. Described here are some of the methods used; a detailed discussion may be found in chapter 8.

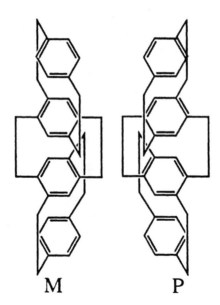

Figure 3.11 M,P system.

Classical Methods

Enantiomers have identical physical and chemical properties. Consequently, they cannot be resolved by conventional methods such as distillation, crystallization, sizing, or chromatography on conventional stationary phases. The resolution of enantiomers in a racemate — a 50:50 mixture of d- and l- forms, requires the use of some external enantionpure or enantio-enriched material or device.

Classically, enantiomers have been separated by forming diastereomeric salts or derivatives with enantio-enriched chiral pool reagents. Since the diastereomeric derivatives thus produced are no longer enantiomers, they can be separated by conventional methods such as crystallization or chromatography on silica gel or other conventional stationary phases. The diastereomers are then de-derivatized to produce purified enantiomers. The methods that have been commonly utilized for generation of optical species are listed here. Chromatographic methods discussed next are generally preferred over other separation methods.

Methods for producing optical species:

- Resolution

    Crystallization
    Chemical separation

- Stereoselective synthesis
- Enzymatic catalysis

Chromatographic Enantioseparation

Enantioseparation on a chiral stationary phase depends on the formation of transient diastereomeric adsorbates with different free energy (see chapter 7 for a more detailed discussion). Stability differences as small as 10 calories can result in detectable HPLC separation. Figure 3.12 illustrates the use of a stationary phase containing an immobilized enantio-enriched selector molecule.

An alternative approach uses an enantio-enriched selector molecule that has been added to the mobile phase, as depicted in Figure 3.13. Table 3.1. lists various chiral stationary phases that can be used for resolving chiral compounds by chromatography.

Enantiopurity

There are a number of ways of determining enantiopurity. The simplest method is polarimetry, discussed next; however, this requires that enantiomeric species have already been resolved.

Polarimetry

The discoveries of polarized light and optical rotation were instrumental in developing the concept of molecular chirality, which, in turn, is basic to better com-

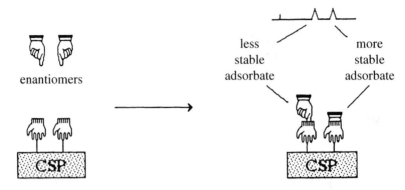

Figure 3.12 Diagrammatic representation of a stationary phase containing an immobilized enantio-enriched selector molecule. Reproduced with permission from S. Ahuja, W. Pirkle, and C. Welch, *Chiral Separations by Chromatography*, American Chemical Society Short Course, 1994. Copyright 1994 American Chemical Society.

prehension of the field of stereochemistry. The methods of "palpitating" chirality by optical tools such as polarimetry, optical rotatory dispersion (ORD), and circular dichroism (CD) have been called chiral-optical methods by Weiss.[16] This term was shortened to chiroptical methods by Vladimir Prelog and a number of other scientists.

When a ray of ordinary light is passed through a specially constructed prism made of Icelandic spar, such as a Nicol prism, the emergent light wave vibration takes place in one plane, and the light is said to be plane-polarized, or simply polarized light. If another Nicol prism is used to examine the polarized light, it is seen

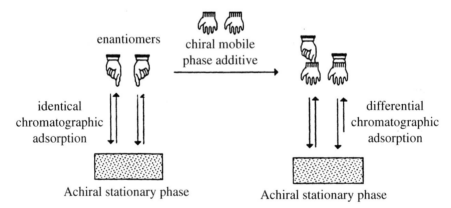

Figure 3.13 Diagrammatic representation of an enantio-enriched selector molecule that has been added to the mobile phase. Reproduced with permission from S. Ahuja, W. Pirkle, and C. Welch, *Chiral Separations by Chromatography*, American Chemical Society Short Course, 1994. Copyright 1994 American Chemical Society.

Table 3.1 Chiral Stationary Phases

Functionality	Basis of Separation	Mode
Amide	Attractive interactions, H bonding, π-interactions	GC, HPLC
Fluoro alcohol	Dipole attraction, charge transfer	HPLC
Carbohydrate	Attractive interactions/inclusion	HPLC
Cyclodextrins	Inclusion/other interactions	GC, HPLC
Crown ether	Inclusion	HPLC
Metal chelates	Ligand exchange	TLC, GC, HPLC
Proteins	Hydrophobic and polar interactions	HPLC

that on rotating the latter prism, the field of view appears alternately light and dark, the minimum brightness following the maximum as the prism is rotated through an angle of 90 degrees. The prism that is used to polarize the light is called a polarizer, and the one that is used to examine the light is called an analyzer. When a tube filled with an optically active solution is placed between the polarizer and the analyzer that are at right angles to each other, the field turns from dark to light because of rotation of the plane of polarization by the optically active substance. The extent to which the plane of polarized light has been rotated can be determined by turning the analyzer through a certain angle until the field becomes dark again. When it is necessary to turn the analyzer to the right (clockwise), the optically active substance is said to be dextrorotatory; when the analyzer has to be turned to the left (counterclockwise), the substance is said to be levorotatory.

The angle of rotation of the polarized light on passage through a substance varies directly with the length of the column of substance through which the light passes. In homogeneous mixtures or solutions, the angle of rotation depends on the concentration of optically active substance present. The angle of rotation also varies with the wavelength of light employed (the shorter the wavelength, the greater the rotation), with the temperature, and with the nature of the substance. When all of these factors are considered, the results may be calculated in terms of specific rotation.

The specific rotation $[\alpha]$ of a liquid is defined as the angular rotation in degrees through which the plane of polarization of polarized monochromatic light is rotated by a passage through 1 decimeter (100 mm) of the liquid, calculated to the basis of specific gravity of 1. In the case of a solution of optically active substance, the specific rotation is calculated to the basis of concentration of 1 g of solute in 1 mL of solution. Since the specific gravity of liquids as well as the optical activity of many substances is influenced by temperature, it is necessary to indicate the temperature at which the rotation and specific gravity have been determined.

The calculation of specific rotation of an optically active solution can be made from the following equation:

$$[\alpha]_D^t = \frac{100a}{lpd}$$

where $[\alpha]_D^t$ = the specific rotation at given wavelength and temperature t. D denotes sodium D-line (589 nm), a = observed rotation in degrees of liquid at temperature t, l = the length of tube in decimeters, p = concentration of substance in solution (g/100 mL), and d = specific gravity of solution at t.

Optical Purity

It is not always necessary to use polarimetric detectors. Optical purity can be determined chromatographically with any usable detector by comparing the sample with an authentic standard. Enantiopurity is generally reported in terms of enantiomeric excess (e.e.).

$$\% \text{ e.e.} = \frac{\text{major} - \text{minor}}{\text{major} + \text{minor}} \times 100$$

If the area of the minor isomer present as an impurity is 1% of the major component, then the % e.e. calculated value is 98% and not 99% pure as some would normally describe it.

The optical purity ($P,\%$) of a given sample with a specific rotation $[\alpha]$ can be determined from the following equation if the specific rotation of an optically pure compound $[\alpha]$max is known:

$$P = 100 \frac{[\alpha]}{[\alpha]\text{max}}$$

This approach to determination of optical purity may be associated with systematic errors and may not always correspond to the actual enantiomeric composition or enantiomeric purity. Optical purity relates to enantiomeric purity only when there is no molecular association between enantiomers. Therefore it is important to use methods that distinguish between enantiomeric forms. In other words, chromatographic or other separation methods may be more useful.

Determination of Absolute Configuration

The stereochemical description (R or S, M or P) and the spatial arrangement of the atoms in a chiral molecule that distinguishes it from its mirror image can be defined as absolute configuration. From a regulatory standpoint, it is necessary to determine the absolute configuration of molecules of therapeutic interest (see chapter 2).

Earlier approaches to determination of absolute configuration depended on the comparison of measured optical rotations with values computed by theory. More recent extensions include the comparison of theoretically computed ORD and CD curves, as well as the excitation chirality method.[2]

The first absolute configuration of any chiral molecule was achieved by Bijovet et al., in 1951.[11] In regular X-ray crystallography, the intensity of diffracted beams depends on the distances between the atoms but not on the absolute spatial orientation of the structure. This occurs because the phase change due to scattering of incident radiation is nearly the same for all atoms. Therefore, a chiral crystal and its enantiomorph produce the same X-ray patterns and cannot be distinguished from one another.

The solution to this problem, in principle, can be found in the studies of Coster et al.,[17] where they used so-called anomalous X-ray scattering to determine the sequence of planes of zinc and sulfur atoms in a crystal of zinc blende (ZnS). The method involves using X rays of a wavelength near the absorption edge of one of the atoms, in this case zinc. This results in a small phase change in X rays scattered by zinc atoms relative to sulfur atoms, which does not depend on their relative positions. As a result, the diffraction pattern is no longer centrosymmetric. Pairs of spots in the pattern that are related by the center of symmetry, which are now called Bijovet pairs, become unequal in intensity. Provided the structure is known except for the absolute configuration, it is possible to calculate the relative intensities of the Bijovet pairs for the R and S isomers, and by comparing calculated values with the experiment data, it is possible to differentiate the isomers. This principle has been applied by Bijovet et al. by using zirconium K_α X rays in the X-ray crystallographic study of sodium rubidium tartrate.

The (+)-tartrate anion has been found to have the R, R configuration. Since (+)-tartaric acid has been correlated with many other chiral compounds, especially the sugars, the determination of its configuration marked a major milestone in stereochemistry.

Neutron diffraction analysis has been found more useful than X-ray analysis for the determination of the absolute configuration of (−)-glycolic-d acid (−)-CHDOH-COOH, obtained by enzymatic reduction of glyoxylic-d acid CDO-COOH with NADH (nicotinamide adenine dinucleotide, reduced form) in the presence of liver alcohol dehydrogenase.[18] The optically active acid is converted to ^{6}Li salt, which is then subjected to X-ray and neutron diffraction analysis. The results are important in that the configuration of numerous other chiral deuterium compounds have been correlated with that of (S)-(−)-glycolic-d acid.

Absolute Configuration and Biological Activity

The example discussed next on determination of absolute configuration of a series of compounds in the pursuit of a potent and selective bronchodilator may be of some interest.[19] The biological properties of β-adrenoreceptor stimulants with the general structure 3-acylamino-4-hydroxy-α-((N-substituted amino)methyl) benzyl alcohol are well known. Like many other members of the series, 3-formamido-4-hydroxy-α[[N-(p-methoxy-α-methylphenethyl)amino]methyl] benzyl alcohol (I) has two asymmetric carbons in its molecule and is a mixture of two pairs of enantiomers IA and IB. These were separated by selective crystallization, and one of them (IA) was found to be highly promising as a potent and selective bronchodilator.

Compounds IA and IB have now been separated further into their respective optical isomers to determine the absolute configurations of these four isomers: (−)-IA, (+)-IA, (−)-IB and (+)-IB. IA, IB and N-(p-methoxy-α-methylphenethyl)-amine (II) were resolved using (+)- and (−)-forms of tartaric acid as resolving agents into their optical isomers as shown in Table 3.2.

Two isomers [(−)-IA and (−)-IB] were also obtained from (−)-II and 4′-benzyloxy-3′-nitro-2-bromoaceto-phenone by several steps. The configurations of the enantiomers of II were determined by chemical correlation with those of N-p-hydroxy-α-methylphenethyl-amine, whose configuration is known. The configurations of the asymmetric carbon (α) carrying the OH group and the asymmetric carbon (β) located in the amine moiety of (+)-IA were respectively determined by correlation with (+)-4-methoxy-3-nitrobenzoic acid and (+)-II, whose configurations are known, through (+)-3-amino-4-methoxy-α-[N-(p-methoxy-α-methylphenethyl)amino]-methylbenzyl alcohol. On the basis of these experiments, the configurations of the four isomers can be depicted as follows: (−)-IA = αR,βR, (+)-IA = αS,βS, (−)-IB = αS,βR, and (+)-IB = αR,βS. Bronchodilator activity of these compounds was found to decrease in the order of (−)-IA > (+)-IB > (+)-IA > (−)-IB.

The bronchodilator activity of the isomers was compared using isolated tracheal preparations of guinea pigs, and the data are given in Table 3.3. Compounds (−)-IA and (+)-IB, which have the R configuration of the α-carbon (there is a β-carbon later), were more potent than the corresponding isomers (−)-IB and (+)-IA, respectively, which have the S configuration. These data are in general agreement with those reported for sympathomimetic amines such as norepinephrine, epinephrine, isoproterenol, and salbutamol, the activity of which is

Table 3.2 Optical Rotation and Melting Point of Isomers of I

Compound	mp (°C)	$[\alpha]_D^2$
(+)-IA	a[a]	+29.3
(+)-IA(+)-tartrate	184	+40.4
(−)-IA	a	+30.1
(−)-IA(−)-tartrate	185	−42.6
(+)-IB	150	+8.9
(+)-IB(+)-tartrate	172	+12.2
(−)-IB	150	−9.0
(−)-IB(−)-tartrate	172	−12.3

a = an amorphous solid.

Note: The rotations of the tartrates were measured in aqueous solutions and those of the bases in methanolic solution.

Source: Reproduced with permission from K. Murose, T. Masc, H. Ida, K. Takahashi, and M. Murakami, Chem. Pharm, Bull,, 26, 1123 (1978). Copyright 1978 Pharmaceutical Society of Japan.

Table 3.3 Potency of the Isomers of I on Tracheobronchial Muscle

Compound	Configurations		Relative Bronchodilator Potencies[a] Dose Ratios (isoproterenol = 1)
	α	β	
Isoproterenol			1.0
Racemic IA[b]			0.1
(−)-IA[b]	R	R	0.08
(+)-IA[b]	S	S	0.31
Racemic IB[b]			0.91
(−)-IB[b]	S	R	1.1
(+)-IB[b]	R	S	0.2

a. Based on the effective dose required to give 50% relaxation of histamine-induced constriction of isolated guinea pig tracheal preparations.

b. The compound was used as its fumarate; the molar ratio of the compound to fumaric acid was 2:1.

Source: Reproduced with permission from K. Murose, T. Mase, H. Ida, K. Takahashi, and M. Murakami, *Chem. Pharm. Bull.*, **26**, 1123 (1978). Copyright 1978 Pharmaceutical Society of Japan.

known to reside almost exclusively in one of their isomers having the R configuration. However, the difference in potency between the isomers of I was far smaller than might have been expected from that of the sympathomimetic amines cited above. (−)-IA with αR,βR configuration was only 14 times more potent than (−)-IB, which has αS,βR configuration, and the difference in potency between (+)-IB and (+)-IA, which have the configuration αR,βS and αS,βS, respectively, was only marginal.

Another interesting feature that may be noted from Table 3.3 is the effect of configuration around the β-carbon on the potency of the isomers. The isomer (−)-IA with αR,βR configuration was about three times as potent as (+)-IB, whose configuration is αR,βS. On the other hand, the potency of (+)-IA with αS,βS configuration was about three times that of (−)-IB, whose configuration is αS,βR. These data show that the configuration at the β-carbon also influences the bronchodilator potency of compound I, although it is impossible to correlate a particular configuration with the increase in potency. In this connection, it is pertinent to quote the β-blocking activity reported for the four isomers of 2-(α-methyl-2-phenethyl-amino)-1-(2-naphthyl)ethanol, which, like compound I, has two asymmetric carbons α and β, the former neighboring the naphthyl group and the latter carrying the amine moiety. The β-blocking activity of the isomers, as reported, decreases in the order αR,βR > αR,βS ≫ αS,βR > αS,βS. The results with this compound show that the R configuration at the β-carbon, as well as at the α-carbon seems to be the preferred configuration for the potency. However, it should be pointed out that the relationship between the stereochemistry and the β-stimulant activity of the series of compounds including I is quite complicated, and much remains to be elucidated before a general rule can be worked out.

REFERENCES

1. S. Ahuja, W. Pirkle, and C. Welch, *Chiral Separations by Chromatography*, American Chemical Society Short Course, 1994.
2. E. L. Eliel and S. H. Wilen, *Stereochemistry of Organic Compounds*. Wiley: New York, 1994.
3. I. W. Wainer, *Drug Stereochemistry*. Marcel Dekker: New York, 1993.
4. S. Allenmark, *Chromatographic Enantioseparation*. Ellis Horwood: New York, 1991.
5. R. J. Haüy, *Traite' de Minerologie*. Chez Louis: Paris, 1801.
6. E. L. Malus, Mem. Soc. d'Árcueil, **2**, 143 (1809).
7. J. B. Biot, Mem. Cl. Sci. Math. Phys. Inst. Imp. Fr., **13**, 1 (1812).
8. J. F. W. Herschel, *Trans. Cambridge Philos. Soc.*, **1**, 43 (1822).
9. L. Pasteur, *Lectures to Societe Chimique de France*, January 20 and February 3, 1860.
10. S. Ahuja, *Chiral Separations: Applications and Technology*. American Chemical Society: Washington, DC, 1996.
11. E. Fischer, *Ber. Dtsch. Chem. Ges.*, **24**, 2683 (1891).
12. J. M. Bijovet, A. P. Peerdeman, and A. J. van Bommel, *Nature*, **168**, 271 (1951).
13. H. Buding, P. Deppisch, H. Musso, and G. Snatzke, *Agnew. Chem.*, **97**, 503 (1985).
14. R. S. Chan, C. K. Ingold, and V. Prelog, *Experentia*, **12**, 81 (1956).
15. R. S. Chan, C. K. Ingold, and V. Prelog, *Angew. Chem. Int. Ed.*, **5**, 385, 511 (1966).
16. U. Weiss, *Experentia*, **24**, 1088 (1968).
17. D. Koster, K. S. Knoll, and J. A. Prins, *Z. Phys.*, **63**, 345 (1930).
18. C. K. Johnson, E. J. Gabe, M. R. Taylor, and I. A. Rose, *J. Am. Chem. Soc.*, **87**, 1802 (1965).
19. K. Murase, T. Masc, H. Ida, K. Takahashi, and M. Murakami, *Chem. Pharm. Bull.*, **26**, 1123 (1978).

REVIEW

Assign stereochemical configuration based on the Cahn-Ingold-Prelog System. See **3.1–3.14** on next page.

TUTORIAL

Assign R or S configuration for the following (Figure 3.14).

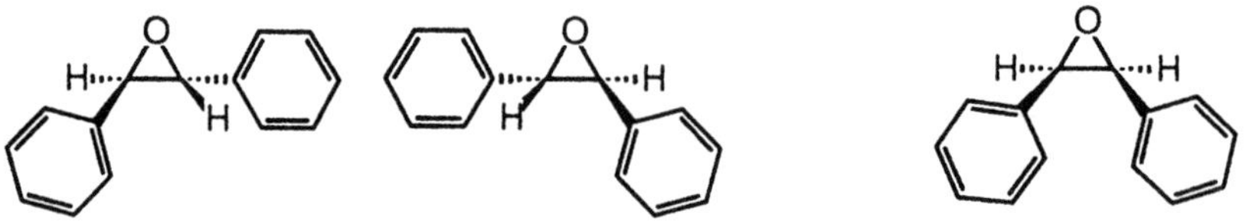

Figure 3.14 Examples of diastereomers. Reproduced with permission from S. Ahuja, W. Pirkle, and C. Welch, *Chiral Separations by Chromatography*, American Chemical Society Short Course, 1994. Copyright 1994 American Chemical Society.

The molecule is stilbene oxide. Its diastereomers are as follows:

1. S,S
2. R,R
3. R,S (meso)

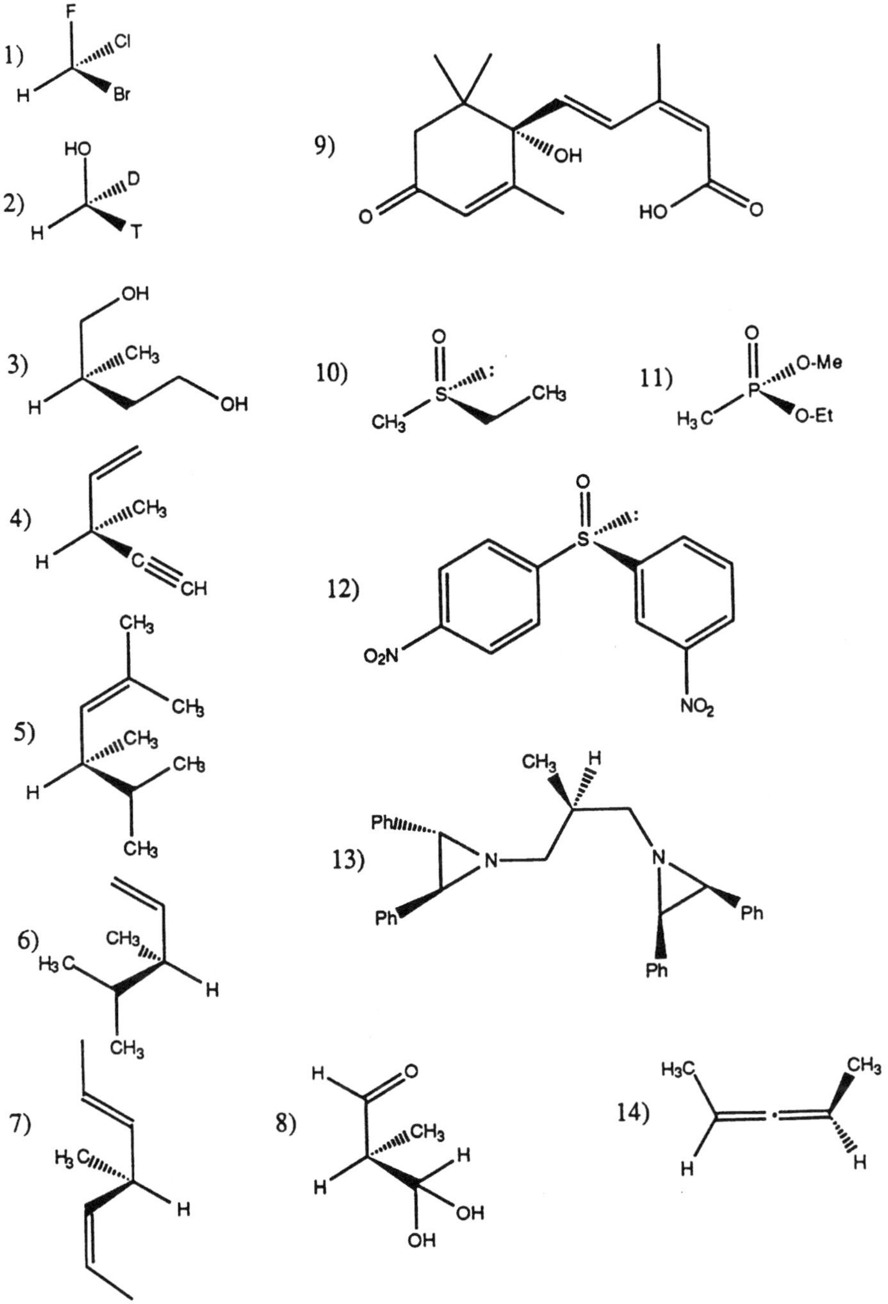

3.1–3.14
Reproduced with permission from S. Ahuja, W. Pirkle, and C. Welch, *Chiral Separations by Chromatography*, American Chemical Society Short Course, 1994. Copyright 1994 American Chemical Society.

4

Separation Methods

The commonly used methods for separation of enantiomers can be broadly classified into the following five categories:

- Thin-layer chromatography (TLC)
- Gas chromatography (GC)
- High pressure liquid chromatography (HPLC)
- Supercritical fluid chromatography (SFC)
- Capillary electrophoresis (CE)

Aside from capillary electrophoresis, all of these methods are based on chromatography. Strictly speaking, capillary electrophoresis is not a chromatographic technique because two phases are not involved in the separation process. It may be recalled that the two phases in chromatography are designated as the stationary phase and the mobile phase, based on their role in the separation process. Technically, there is no stationary phase in capillary electrophoresis unless the capillary walls are assigned that role. Some chromatographers promote this concept, but it is not entirely correct. In any event, most chromatographers are comfortable in using CE because some of the manipulations used to optimize chromatographic separations are also suitable for CE. And symposia on CE are often included in the major chromatographic meetings. CE is included here because of its similarity to chromatography and the fact that it is a very useful technique for enantiomeric separations.

The following discussion is limited to introduction of these techniques and how they are best utilized for enantiomeric separations. All of these methods except TLC are also discussed in later chapters in this book, and their applications may be seen in chapter 10.

Thin-Layer Chromatography

TLC is a very useful qualitative technique that entails minimal costs. It can provide good indications as to which method would be best suited for resolving enantiomers. Of course, it can be used as an independent technique with limitations of resolution and low precision. A significant amount of coverage is provided here to enable the reader to try TLC. Additional reference sources are also provided for TLC aficionados.

As already mentioned, in all chromatographic methods including TLC, it is necessary to have two phases to achieve a successful separation. These phases are designated as the stationary phase and the mobile phase. In conventional TLC, the stationary phase is generally silica gel, and the mobile phase is composed of a mixture of solvents. Contrary to common belief, the observed separation in TLC is not due to adsorption on silica gel alone. There is always a finite amount of water present in the silica gel plates, which acts as a partitioning agent. Furthermore, the mobile phase solvents are adsorbed onto silica during development and thus provide yet another mechanism for partition. Ionic sites in silica gel permit ion exchange, and metallic impurities provide mixed mechanisms for certain separations.

The approaches to separations in chiral chromatography can be classified as follows:

- Achiral stationary phase and achiral mobile phase
- Achiral stationary phase and chiral stationary phase additives
- Chiral stationary phase (CSP) and achiral mobile phase

Achiral Stationary Phase and Achiral Mobile Phase

To utilize an achiral stationary phase and an achiral mobile phase, derivatization is required to separate the diastereomers. (−)-Menthyl chloroformate or (S)-camphor-10-sulfonyl derivatives are commonly used. Because of the difficulty of preparation of derivatives and the attendant problems of by-products, this technique is not as popular as it should be. A few examples of stereoisomeric separations with several derivatization reagents are given in Table 4.1 for a number of interesting pharmaceutical compounds. It may be noticed that all of the separations given in the table use silica gel plates with or without a fluorescent indicator. The detection can then be easily performed under UV light at 254 nm.

Achiral Stationary Phase and Chiral Stationary Phase Additives

A number of chiral eluent additives have been investigated. These include the following compounds:

- β-cyclodextrin
- D-galacturonic acid

Table 4.1 Stereoisomeric Separations

Compound	Rf (R,S)	Eluent	Derivatization Reagent	Plate
Amphetamine	0.21, 0.14	Toluene:dichloromethane:tetrahydrofuran (5:1:1)	(S)-(+)-Benoxaprofen chloride	Silica gel 60
Methamphetamine	0.33, 0.27	as above	as above	as above
Methylbenzylamine	0.28, 0.16	as above	as above	as above
Naproxen	0.53, 0.63	Chloroform:ethanol:acetic acid (9:1:0.5)	(1R, 2R)-(−)-1-(4-Nitrophenyl)-2-amino-1,3-propanediol	Silica gel F 254
Metoprolol	0.24, 0.28	Toluene:acetone (100:10)	(S)-(+)-benoxaprofen chloride	Silica gel 60
Oxprenolol	0.32, 0.38	as above	as above	as above
Propranolol	0.32, 0.39	as above	as above	as above

Source: Adapted from K. Gunther, *Handbook of Thin Layer Chromatography*, Sherma and Fried, Eds., Marcel Dekker, New York, 1991.

- (R)-N-(3,5-dinitrobenzoyl) phenylglycine
- N-(1R, 3R)-*trans*-chrysanthemoyl-L-valine
- (+)-Tartaric acid
- (−)-Brucine

An example using β-cyclodextrin for separation of derivatized amino acids is discussed here because cyclodextrins have been commonly used in separations of chiral compounds.

It may be seen from the structure of β-cyclodextrin, given in Figure 4.1, that β-cyclodextrin is a chiral toroidal-shaped molecule with a finite cavity formed by the connections of seven glucose units via 1,4-linkage. Retention is related to the size of the cavity of the cyclodextrin and other attendant interactions of the enantiomers with the oligomer.

2-Naphthylamide and *p*-nitroanilide derivatives of amino acids have been separated on Sil C18.50F plates with the following mobile phase containing β-cyclodextrin:[1] aqueous solution (100mL) containing sodium chloride (2.5 g), urea (26 g), acetonitrile (20 mL), and 0.15 M β-cyclodextrin.

The plate is developed to a distance of 7 cm (Figure 4.2). The following racemates can be resolved: DL-methionine-2-naphthylamide (spot #2), DL-leucine-2-naphthylamide (spot m_1), and DL-leucine-*p*-nitroanilide (spot m_2). D and L alanine-*p*-nitroanilides are *not* resolved by this method.

Chiral Stationary Phase and Achiral Mobile Phase

For enantiomeric separations, the TLC plates can be coated with the following materials to provide useful stationary phases: cellulose, chiral compounds, cyclodextrin, and ligand exchange. Commercial plates have also been used.

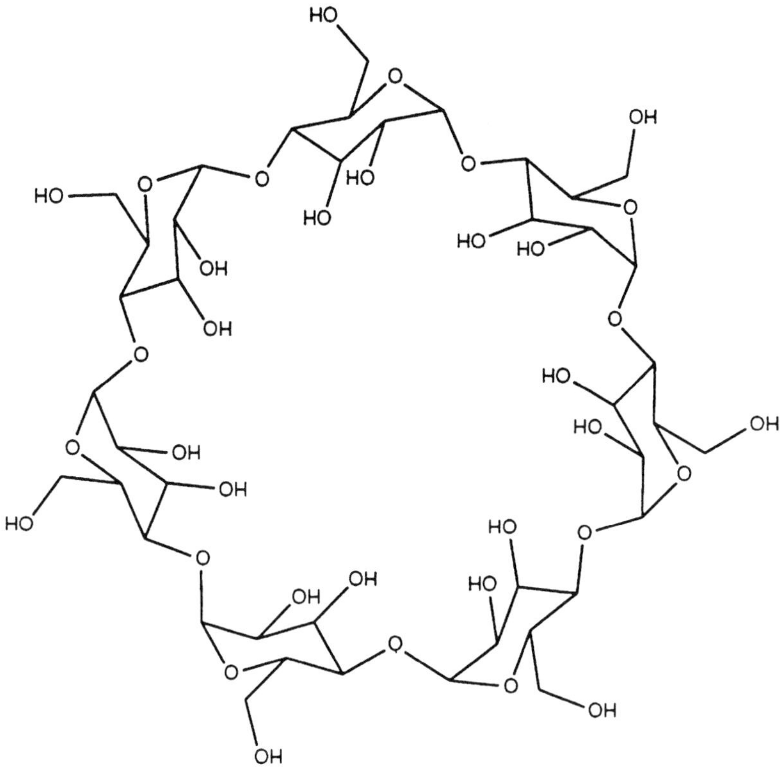

Figure 4.1 Structure of β-cyclodextrin.

Cellulose plates. Cellulose is a linear macromolecule composed of optically active D-glucose units with helical cavities (Figure 4.3). The separation of enantiomers is influenced by the way they fit differently in the lamellar chiral layer structure of the support. Peracetylation of the cellulose is performed in a way that assures that conformation and relative position of the carbohydrate bands in their crystalline domain remain intact. Systematic investigations of this chiral support has resulted in commercialization of a microcrystalline triacetylcellulose plate. These plates are stable with aqueous eluent systems and resistant to dilute acids and bases. They are also stable in alcoholic eluents, but are attacked by glacial acetic acid and ketonic solvents.

Chiral compounds plates. Described here are a few interesting examples of separations of a number of amino acids by TLC on silica gel/(+)-tartaric acid plates. PTH derivatives of amino acids have been resolved on silica gel/(+)-tartaric acid plates.[2] The mobile phase is composed of a mixture of chloroform : ethyl acetate : water (28:1:1), and the plates were developed for a distance of 10 cm. The hRf values of various amino acids resolved are given in Table 4.2.

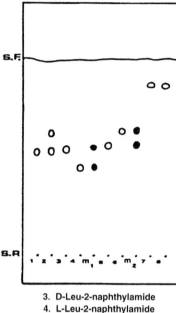

3. D-Leu-2-naphthylamide
4. L-Leu-2-naphthylamide
M,-mixture of 3,4

Figure 4.2 Resolution of methionine and leucine. Reproduced with permission from L. Lepri, V. Ceas, and P. G. Deideri, *J. Planar Chromatography*, **4**, 338 (1991).

The resolution of enantiomers of various amino acids is very good; however, the best resolution of D- and L-isomer is obtained for tyrosine.

β-Cyclodextrin plates. A number of derivatized amino acids as dansyl or naphthylamide derivatives have been chromatographed on β-cyclodextrin plates with varying ratios of methanol and 1% triethylammonium acetate.[3] Good resolution is generally obtained between the D- and L-derivative of the amino acid at pH 4.1 (Table 4.3). Detection can be performed by fluorescence or spraying with ninhydrin reagent.

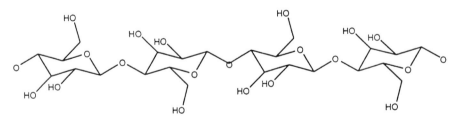

Figure 4.3   Structure of cellulose.

Table 4.2 hRf Values of Resolved Enantiomers of PTH Amino Acids

DL Mixture	hRf of D	hRf of L
Methionine	16	83
Phenyl alanine	15	85
Valine	21	80
Tyrosine	16	95
Threonine	30	85
Alanine	12	55
Serine	10	84

Source: Data from L. Lepri, V. Ceas, and P. G. Deideri, *J. Planar Chromatogr.*, 4, 338 (1991).

Ligand exchange plates. Aromatic amino acids have been separated on Merck RP-18 WF 254 S plates covered with copper acetate and LNDH (L-N-n-decylhistidine). The separation data are shown in Table 4.4. The mobile phase is composed of methanol:ACN:THF:water in the ratio of 7.3:5.9:33.9:52.9[4] and α values for the enantiomers range from 1.19 to 1.43.

Commercial plates. Chiralplate, the first ready-to-use TLC plate, was marketed by Macherey-Nagel. This was followed by the chiral, HPTLC plate, CHIR, in 1988, which further simplifies the use of TLC for chiral separations. In 1993, Astec announced availability of β-cyclodextrin plates for scouting chiral separations prior to HPLC.

Gas Chromatography

Again, using gas chromatography (GC) for analysis of those compounds that lack volatility and have inherent thermal instability is difficult because derivatization,

Table 4.3 TLC of Amino Acids on β-Cyclodextrin Plates

Compound	Rf D	Rf L	Mobile phase	Detection
Dansyl leucine	0.49	0.66	40/60	Fluorescence
Dansyl methionine	0.28	0.43	25/75	Fluorescence
Dansyl alanine	0.25	0.33	25/75	Fluorescence
Dansyl valine	0.31	0.42	25/75	Fluorescence
Alanine-β-naphthylamide	0.16	0.25	30/70	Ninhydrin
Methionine-β-naphthylamide	0.16	0.24	30/70	Ninhydrin

Source: Reproduced from A. Alak and D. W. Armstrong, *Anal. Chem.*, 58, 584 (1986). Copyright 1986 American Chemical Society.

Table 4.4 TLC Resolution of Aromatic Amino Acids by Ligand Exchange

Amino Acid	Rm (L)	Rm (D)	α
Tryptophan	0.50	0.66	1.33
α-Methyl-tryptophan	0.56	0.73	1.38
5-Methyltryptophan	0.60	0.75	1.33
6-Methyltryptophan	0.63	0.72	1.19
Phenylalanine	0.07	0.25	1.28
α-Methylphenyl-alanine	0.18	0.42	1.43
Tyrosine	−0.03	0.13	1.21
α-Methyl tyrosine	0.06	0.24	1.27
DOPA [3-(3,4-dihydroxyphenyl) alanine]	0.08	0.24	1.24

Source: Reproduced with permission from S. Allenmark, *Chromatographic Enantioseparation*, Ellis Harwood, New York, 1991. Copyright 1991 *Chromatographia*.

with its attendant problems, is necessitated. Development of columns with better thermal stability and advances in capillary column technology have generated a renewed interest in this technique. GC offers advantages such as high resolution and large peak capacity. Furthermore, it is useful for resolution of nonaromatic compounds used in asymmetric synthesis, which are not easily separated by TLC or high pressure liquid chromatography (HPLC).

Let us review the problems encountered in performing gas chromatography of enantiomers:

1. The sample must be made volatile if it is not already volatile.
2. The analyte can racemize under chromatographic conditions.
3. Racemization of stationary phases can also occur at high temperatures.
4. Preparative-scale separations are generally achieved with great difficulty.

The following approaches have been used for separation of enantiomers by GC: (a) formation of diastereoisomers that can be resolved on conventional columns and (b) resolution of enantiomers on chiral stationary phases with or without derivatization of the analyte.

Derivatization

Preparation of suitable derivatives for GC allows the use of conventional columns. Table 4.5 gives some derivatization reagents that can be used for alcohols, carboxylic acids including amino acids and hydroxy acids, and several amino compounds (for example, amines, amino acid esters, and amino alcohols).

Some of the disadvantages of derivatization are the following: the chiral reagent has to be enantiomerically pure; the resulting diastereomer should be volatile and stable; and racemization is possible.

Achiral reagents can be used to improve volatility. They can help enhance stereoselective interactions. Furthermore, racemization is avoided as well as pro-

Table 4.5 Derivatization Reagents

Compound Derivatization	Reagent
Amino acids	(+)-3-methyl-2-α-butanol
Carboxylic acids	(+)-3-methyl-2-α-butanol
Hydroxy acids	(+)-3-methyl-2-α-butanol
Amines	L-chloroisovaleryl chloride
Amino alcohols	L-chloroisovaleryl chloride
Amino acid esters	L-chloroisovaleryl chloride
Alcohols	L-chloroisovaleryl chloride

duction of undesirable by-products. Table 4.6 provides derivatization reagents for a variety of compounds. It also shows the resultant products of interest.

Chiral Stationary Phases

The discovery of useful CSPs for GC is fraught with complications, as indicated by the following properties: thermal properties of CSP should be adequate; stereochemical structure should allow chiral discrimination; and column efficiency should be high.

As a general rule, an improvement in enantiomeric separation factor α is obtained by decreasing column temperature. The reason relates to the difference in enthalpy of association of enantiomers with the stationary phase at a given temperature — a difference that decreases with temperature. This means that derivatives should be as volatile as possible.

The separations of chiral compounds on chiral stationary phases are usually carried out on three types of chiral stationary phases: amino acid derivatives, cyclodextrin derivatives, and chiral metal complexes.

Amino acid derivatives. Gil-Av et al.[5] were the first to utilize N-trifluoroacetyl-L-isoleucine lauryl ester phase to separate N-trifluoroacetyl (TFA) derivatives

Table 4.6 Derivatization Methods

Compound	Product	Derivatization Reagent
Alcohols	Isocyanates	Urethanes
Carbohydrates	Phosgene	Carbonates
	Acylating reagents	Acylated products
Amines	Acylating reagents	Acylated product
Amino acids	As above	As above
Amino alcohols	Phosgene	Oxazolidinones
Hydroxy acids	Isocyanates	Urethanes, amides
Carboxy acids	Phosgene	1,3-dioxolane-2,4-diones
Ketones	Hydroxylamine HCl	Oximes

of amino acids. They used a glass capillary column of 100 meters in length and found that D-enantiomers elute first. They attributed the separation to hydrogen bonding, which they assumed occurred between the amide group of CSP and carbonyl oxygen atoms of the solute. In this context, it may be instructive to look at significantly different α values and $-\Delta\Delta G$ values offered by the two selectands consisting of two different esters of N-TFA leucine (Table 4.7).

The volatility and stability of the stationary phase is a common problem in gas chromatography. This situation can be substantially improved by the use of dipeptide phases. Furthermore, the chiral discrimination is increased to permit the use of small packed columns. This helps reduce the column length to 2 meters from 100 meters used by Gil-Av et al.

Other gas chromatographic phases of interest in this group are modified diamides, which have the following long-chain alkyl groups: *n*-dodecyl-L-valine-*t*-butylamide and *n*-decanoyl-(S)-α-(1-naphthyl) ethylamide

An example of a naturally occurring mode of hydrogen bond association leading to separation of enantiomers has been reported by Feibush et al.[6] They studied the resolution of pharmaceuticals such as hydantoins, barbiturates, and glutaramides that contain an $-OC-NH-CO-$ moiety in the molecule. The authors linked N,N'-2,6-diaminopyridinediylbis[S-2-phenylbutanamide] to silica gel through an *n*-undecyloxy handle to produce a CSP that showed selectivity for the above-mentioned compounds. On the basis of their studies, it can be shown that the S selectand of the barbiturate fits better than the R isomer, as would be expected from the order of elution (Figure 4.4).The two large substituents at the 1-position of the S selectand (cyclohexyl) and the α-carbon of the amide group of the selector (phenyl) are *trans* to each other.

A polysiloxane-based polymer with a chiral side chain is produced by copolymerization of dimethylsiloxane with (2-carboxypropyl) methoxy silane and the coupling with L-valine-*t*-butylamide. Polysiloxane-L-valine-*tert*-butylamide has been found to very useful and is sold commercially as Chirasil-Val (**4.1**), a series of coated capillary columns.

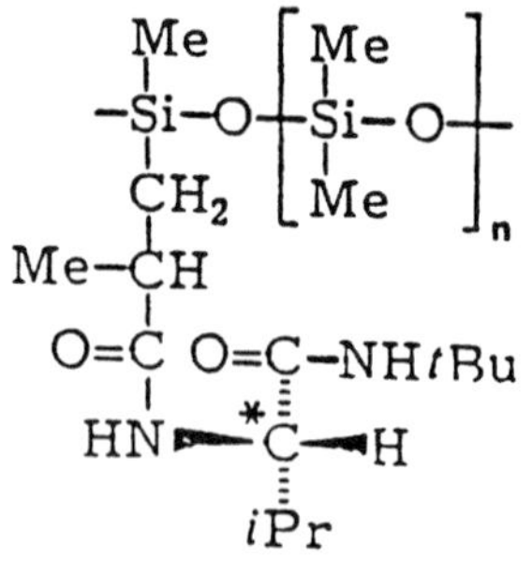

4.1

Table 4.7 Resolution Coefficients (α) for Different Types of Selector/Selectand Systems

	Selector	Selectand	α (temp)	$-\Delta\Delta G$ (cal)[a]
(I)	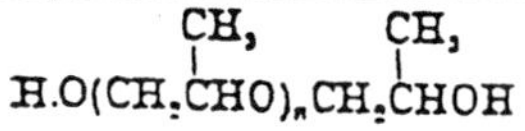 CH₃ CH₃ H.O(CH₂CHO)ₙCH₂CHOH	Norbornanols	1.01 (70°C) GC	7
(II)	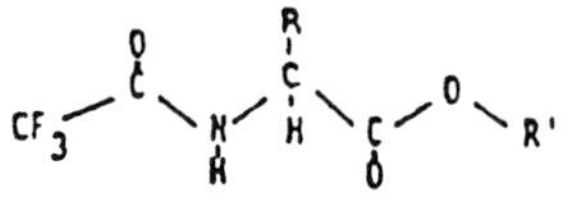 CF₃–C(O)–N(R)–CH(R)–C(O)–O–R'	N-TFA-Leu t-Bu Ester	1.08 (90°C) GC	55
(III)	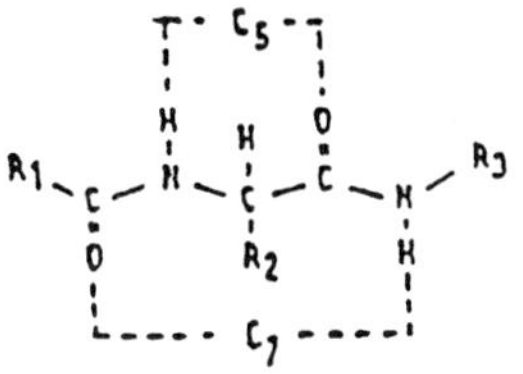 (cyclic peptide structure with R₁, R₂, R₃, C₅, C₇)	N-TFA-Leu iPr Ester III: R₁ = n – C₁₁H₂₃; R₂: iPr, R₃ = t – Bu	1.34 130°C [3.5–4.0] (extrapolated to 0°C) GC	235 760
(IV)	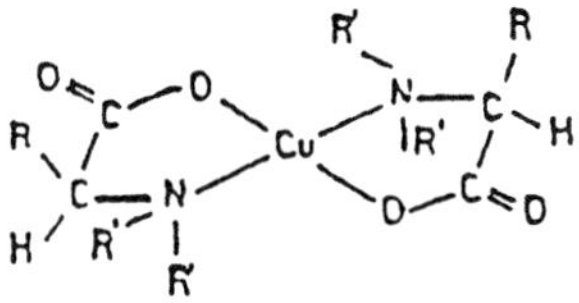 (Cu complex structure)	Valine	4.8 (0°C) HPLC	860
(V)	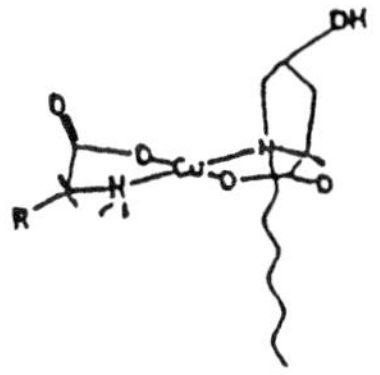 (chiral stationary phase structure)	Proline	4.8 (20°C) HPLC	1400
(VI)	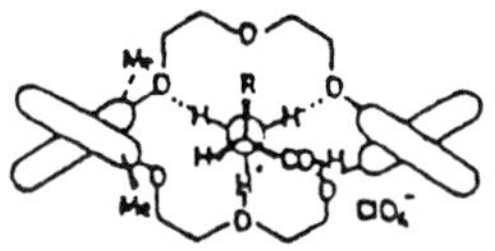 (crown ether structure)	[Phe]+/ClO₄⁻	30* (0°C)	1860
(VII)	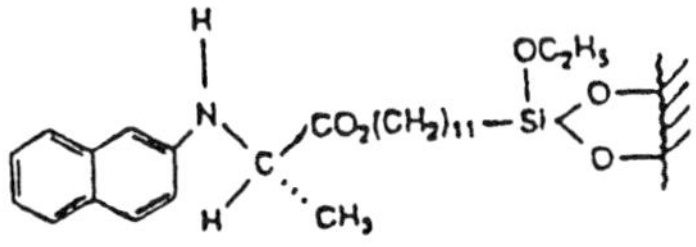 (naphthyl-based CSP structure)	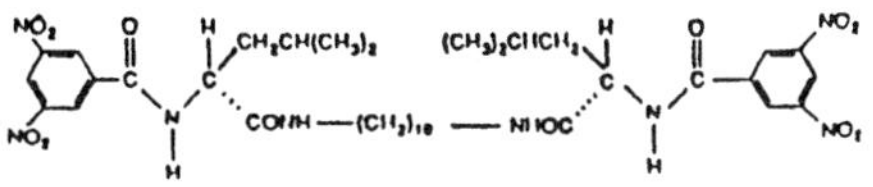	121 (25°C) HPLC	2860

Source: Reproduced from E. Gil-Av, B. Feibush, and R. Charles, ACS Symposium Series #471, S. Ahuja, Ed., American Chemical Society, Washington, D.C., 1991. Copyright 1991 American Chemical Society.

Note: Enamtiomer distribution was constant between two immiscible solvents.

a. $-\Delta\Delta G = RT \ln \alpha$.

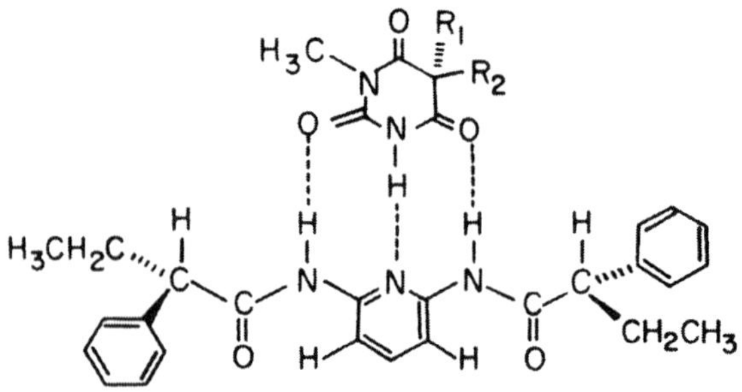

Figure 4.4 Interactions of solute and CSP. Reproduced with permission from B. Feibush, A. Figueroa, R. Charles, K. D. Onan, P. Feibush, and B. L. Karger, *J. Am. Chem. Soc.*, **108**, 3310 (1986). Copyright American Chemical Society.

This column exhibits good temperature stability and can be used up to 220°C. It has been found useful for chromatographing perfluoroacylated and esterified amino acids and amino alcohols.

Calculations show that the L-valine amides adopt a β-pleated sheet or preferentially an (R)-α-helical structure.[7] The energetically favored conformation in a bonding situation will be also dependent on the ligand. It has been assumed on the basis of experimental as well as theoretical results that the immobilized chiral selector of Chirasil-Val will form association complexes by multiple hydrogen bonding interaction, either by an intercalation mechanism or by complexation in an α-helical conformation. Obviously, this model does not apply to ligands that lack donor groups. Examples of compounds that have been resolved on Chirasil-Val are diester and dicarbonyl compounds. A relevant case in this connection is complete separation of atropisomeric 2,2'-binaphthol dipentafluoropropionates, where the (R)-form elutes before the (S)-enantiomer. This suggests that a less favorable conformation has to be adopted for the chiral selector when it binds to the (R)-enantiomer of the ester to allow the bound L-valine-*tert*-butylamide selector to utilize both NH donors for hydrogen bonding to the ester carbonyl groups in the substrate.

A large number of amino acids have been separated on cross-linked CSP (polycyanoethylvinyl siloxane-L-valine-$NHC(CH_3)_3$) by temperature programming up to 190°C (Figure 4.5). The D-enantiomers elute first.[8]

Cyanopropyl silicones have been transformed into suitable materials for chiral derivatization so as to introduce some new chiral ligands. The experiments entail hydrolysis of the cyano group as well as their reduction to form primary amines, which permits the attachment of optically active acids. An interesting phase originating from these studies is L-valine-(R)-1-phenylethylamide coupled to a polysiloxane, which has been shown to be useful for optical resolution of a variety of racemates including O-TFA derivatives of carbohydrates.

Cyclodextrins (CD). In 1983, the first enantiomeric separation using an inclusion-type CSP in gas chromatography was reported for α- and β-pinene and *cis*- and

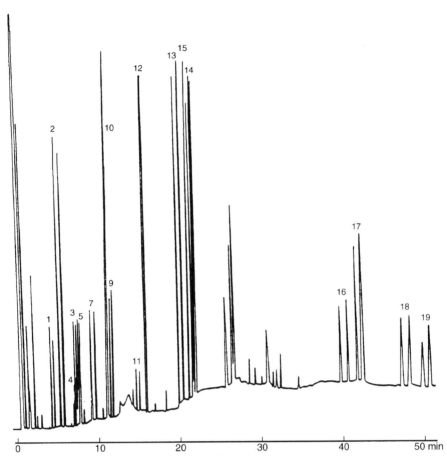

Figure 4.5 Enantiomeric separation of amino acids by GC. Reproduced with permission from X. Lou, Y. Kiu, and L. Zhou, *J. Chromatography*, 552, **153** (1991).

trans-pinane on packed columns containing native α-cyclodextrin in formamide.[9] As mentioned, these compounds are chiral, toroidal-shaped molecules that are composed of six to eight 1,4 linked glucose units: α-CD has six units, β-CD has seven units, and γ-CD has eight units. They are crystalline, thermally stable, and insoluble in most organic solvents. They can be coated on a solid support such as Celite to serve as chiral selectors. A remarkable enantioselectivity can be obtained for certain hydrocarbons; however, the GC column has to be used at low temperatures (<70°C).

Alkylated cyclodextrins can be employed in high-resolution capillary columns for enantiomeric analysis by using neat permethylated β-cyclodextrin [heptakis (2,3,6-tri-O-methyl)-β-cyclodextrin] by using it above its melting point and down to 76°C.[10] Per-*n*-pentylated and 3-acyl-2,6-*n*-pentylated CDs are viscous liquids at room temperature. Konig and coworkers[11] used the following CD derivatives in undiluted forms for the separation of many enantiomeric classes of compounds on deactivated Pyrex glass capillary columns:

- Hexakis (2,3,6-tri-O-n-pentyl)-α-cyclodextrin
- Hexakis (3-O-acetyl-2,6-di-O-n-pentyl)-α-cyclodextrin
- Heptakis (2,3,6-tri-O-n-pentyl)-β-cyclodextrin
- Heptakis (3-O-acetyl-2,6-di-O-n-pentyl)-β-cyclodextrin
- Octanis (2,3,6-tri-O-n-pentyl)-γ-cyclodextrin
- Octanis (3-O-butanoyl-2,6-di-O-n-pentyl)-γ-cyclodextrin

Konig et al. are credited with derivatizing hydroxyl groups of cyclodextrin by introducing alkyl or alkyl-acyl substituents to produce derivatized CDs with low melting points. These CDs are stable and can be used to coat GC columns. For example, perpentylated β-CD can resolve TFA derivatives of sugars, amino acids, and amino alcohols.

The reactivity of 2-, 3-, and 6-hydroxyl groups of glucosidic units is different, so it is possible to modify CD with different substituents. For example, more polar permethyl-O-[(S)-2-hydroxypropyl]-CD offers opposite enantioselectivity from analogous nonpolar alkyl–derivatized CD.

More polar CD derivatives containing hydroxypropyl, free hydroxy, or trifluoroacetyl groups were developed by Armstrong and associates for fused silica capillary columns:[12]

- Hexakis (2-O-(S)-2-hydroxypropyl-3,6-di-O-methyl)-α-cyclodextrin (PMHP-α-cyclodextrin)
- Heptakis (2-O-(S)-2-hydroxypropyl-3,6-di-O-methyl)-β-cyclodextrin (PMHP-β-cyclodextrin)
- Hexakis (2,6-di-O-n-pentyl)-α-cyclodextrin (Dipentyl-α-CD)
- Heptakis (2,6-di-O-n-pentyl)-β-cyclodextrin (Dipentyl-β-CD)
- Heptakis (3-O-trifluoroacetyl-2,6-di-O-n-pentyl)-β-cyclodextrin (DPTFA-β-CD)

Schurig and Nowotny[13] combined the enantioselectivity of CDs with the excellent coating properties and efficiency of polysiloxanes by dissolving alkylated CDs in moderately polar silicones such as OV-1701. This allowed the use of CD derivatives for gas chromatographic enantiomeric separations regardless of their melting points and phase transitions: Heptakis (2,3,6-tri-O-methyl)-β-cyclodextrin in OV-1701 (Chromapack) and Heptakis (2,3,6-tri-O-trifluoroacetyl)-β-cyclodextrin.

These phases show a very broad range of applications for various classes of compounds. For example, Schurig et al.[14] have reported resolution of racemic isomenthol, menthol, and neomenthol at 85°C on 25 m × 0.25 mm i.d. columns coated with heptakis (2,3,6-tri-O-methyl)-β-cyclodextrin on OV 1701.

A number of commercially available CD phases available from Astec are listed here. It may be noticed that the stationary phases consist of various derivatives of α, β, and γ cyclodextrins. Astec also markets Chiraldex columns, which contain cyclodextrin-based phases for chiral separations by capillary gas chromatography. Machrey Nagel sells a number of stationary phases called Lipodex A, B, C, and D based on various substituents on α- or β-cyclodextrins. A large variety of applications by these and other manufacturers of columns are discussed in chapter 10.

- 2,6-di-O-pentyl-α-cyclodextrin
- 2,6-di-O-pentyl-β-cyclodextrin
- 2,6-di-O-pentyl-γ-cyclodextrin
- permethyl-(S)-2-hydroxypropyl-α-cyclodextrin
- permethyl-(S)-2-hydroxypropyl-β-cyclodextrin
- permethyl-(S)-2-hydroxypropyl-γ-cyclodextrin
- 2,6-di-O-pentyl-3-O-trifluoroacetyl-α-cyclodextrin
- 2,6-di-O-pentyl-3-O-trifluoroacetyl-β-cyclodextrin
- 2,6-di-O-pentyl-3-O-trifluoroacetyl-γ-cyclodextrin

A few examples of chiral separations are discussed here. Ethosuximide has been resolved on a 30 m × 0.25 mm i.d. β-cyclodextrin column with a film thickness of 0.25 μm. The run was conducted isothermally at 160°C with a flame ionization detector.[15] The resolution was obtained in fewer than 9 minutes (Figure 4.6).

Separation of the enantiomers of hexobarbital and mephobarbital has been also reported on this column at 250/300°C.[15] The resolution of both compounds was obtained in 17 minutes in the same run (Figure 4.7).

Variously substituted tetralins and indans were separated on different functionalized cyclodextrins as shown in Table 4.8.[16] The column temperature ranged from 70 to 120°C, and the alpha values ranged from 1.01 to 1.06.

Chiral metal complexes. A new technique, complexation gas chromatography, based on the principles of metal coordination, was introduced in 1977.[17] It was demonstrated that a chiral metal coordination compound dicarbonylrhodium(I)-3-trifluoroacetyl-(1R)-camphorate for resolution of enantiomers of 3-methylcyclopentene is useful for separation of enantiomers of 3-methylcyclopentene. The metal complex is used as a solution in squalane ($C_{30}H_{62}$) and coated onto a capillary column. Satisfactory thermal stability and low volatility of the complex and hydrocarbon make such columns useful in the temperature range from below room temperature to 100°C. Since the enantioselective contribution to solute retention is based on electron donation to the metal, the method is well suited to relatively nonpolar compounds with π- or lone-pair electrons, such as cyclic alkenes, ethers, thioethers, and ketones. It is also useful for donors such as alcohols and aziridines.

Structures of some chiral metal chelates are shown in **4.2**.

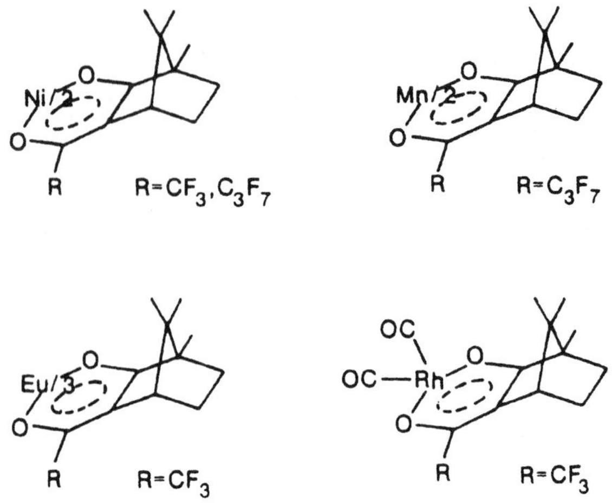

4.2

A limiting factor of the coordination-type CSPs is the low temperature range of operation, typically 25–120°C. The thermal stability has been significantly improved by preparation of immobilized polymeric CSPs (Chirasil Metal). Temperatures as high as 115°C have been used to provide resolution of cyclic ethers in less than 20 seconds on immobilized Chirasil-Nickel.[17] This CSP has also been used for complexation SFC.

Mode Selection in Gas Chromatography

A common feature of all GC phases based on hydrogen bonding is that they are suitable for compounds containing polar functional groups, such as amides, esters, and alcohols. The separation factor generally increases with decreasing column temperature.

CSPs based on metal complexation are suitable for compounds that are less polar and are consequently more volatile. Metal complexation abilities are found in simple alkenes as well as other compounds with electron-donating orbitals, such as esters, ethers, and thioethers; many of these compounds do not require derivatization. These columns can be operated at low temperatures. Interesting applications of these columns have been found in head space analysis of chiral alkenes or pheromones.

The chiral selectors that may form inclusion complexes with the analytes are highly versatile and are used commonly for GC of chiral compounds (see chapter 10). The columns containing α-cyclodextrin dissolved in stationary phase are useful for separating saturated hydrocarbons. Alkylated cyclodextrin phases show enantioselectivity for a wide variety of compounds.

Column:	Cyclodex-B ™ 30m x 0.25mm I.D. J&W P N 112-2532
Film Thickness:	0.25 micron
Oven:	160°C Isothermal
Carrier:	Hydrogen @ 1.10 mL/min (37 cm/sec)
Injector:	Split 1:90; 250°C 1 µL of 5 µg/µL in methanol
Detector:	FID: 300°C Nitrogen make-up gas @ 30 mL/min

(±)- Ethosuximide

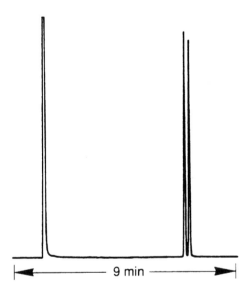

Figure 4.6 Resolution of ethosuximide by GC. Reproduced with permission from J & W Scientific.

Capillary gas chromatography may be useful since the separation factor α is generally low with CD derivatives. The film thickness commonly employed is 0.25 µm, and it is important to remember that α values are temperature-dependent — low temperature is to be favored, where possible. The potential user is well advised to review published chromatograms because they give the best clue to parameters such as selectivity, efficiency, capacity factor, retention, baseline stability, and detectability. Chiral test mixtures should be used to compare columns from various sources.

Since a very wide range of classes of compounds can be resolved on CD derivatives, this type of chiral selector should be considered first for resolving an enantiomeric separation. If peak switching of the enantiomers is desired, then the hydrogen bonding or coordination-type chiral selectors should be considered.

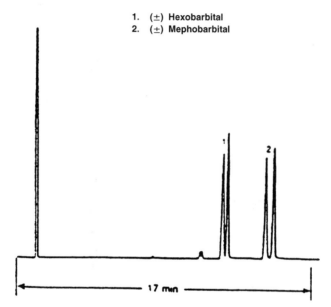

Figure 4.7 Resolution of barbiturates by GC. Reproduced with permission from J & W Scientific.

High Pressure Liquid Chromatography

Discussion in chapters 6–8 covers at length various examples of separation methods that use HPLC. Here, the discussion is limited to various ways that are available to resolve enantiomers by HPLC.

Chromatography of Diastereomeric Derivatives

Precolumn derivatization of an optically active solute is carried out with another optically active molecule. This approach is useful for compounds containing amino groups, hydroxyl groups, or carboxyl groups. Epoxides, olefins, and thiols can also be resolved. A number of examples are discussed later in this book.

Table 4.8 GC Separation of Hydroaromatic Biomarkers

Name	k′a	α	Temp. (°C)	Columnb
Tetralin				
1,8-di-methyl-	6.9	1.03	110	10 m.B
2,7-di-methyl-	38.4	1.02	70	10 m.B
	21.1	1.05	90	10 m.A
1,5,8-tri-methyl-	10.3	1.05	120	10 m.B
2,6-di-methyl-	35.2	1.03	100	10 m.A
	12.42	1.04		
1,4-di-methyl-	21.5	1.06c	80	10 m.B
	58.4	1.01c	70	10 m.G
2-ethyl-	40.5	1.01	70	10 m.B
	12.9	1.02	100	10 m.A
Indan				
1-isopropyl-	19.4	1.02	90	10 m.A
1-propyl-	19.4	1.02	90	10 m.A
1-ethyl-	15.0	1.05	90	10 m.A

a. The k′ value is for the first eluted enantiomer.

b. G-2,6-d (O-pentyl-3-O-trifluoroacetyl)-γ-cyclodextrin; A & B-Permethyl derivatives of O-((s)(2-hydroxypropyl) cyclodextrin.

c. This compound exists as a pair of enantiomers and a meso compound. This α value is for the enantiomeric pair only.

Source: Reproduced from ref. 16, Copyright 1991 American Chemical Society.

Resolution Using Chiral Mobile-Phase Additives

This depends on formation of diastereomeric complexes with a chiral molecule added to the mobile phase. Chiral resolution is due to the differences in the stabilities of diastereomeric complexes of various molecules, solvation in the mobile phase, or binding of the complexes to the solid support. The type of separations can be based on transition metal complexes, ion pairs, or inclusion complexes. Examples of separations by these mechanisms may be found in chapter 8.

Resolution on Chiral Stationary Phases

Enantiomers can be resolved by formation of diastereomeric complexes between the solute and a chiral molecule that is bound to the stationary phase. Separation of enantiomeric compounds on CSP results from the differences in energy between temporary diastereomeric complexes formed between the solute isomer and the CSP; the larger the difference, the greater the separation. The observed retention and efficiency of a CSP, however, is based on the total of all interactions between the solute isomer and the CSP, including achiral interactions. A large number of such separations are included in this book (see chapters 6–8 and 10).

Supercritical Fluid Chromatography

In supercritical fluid chromatography (SFC), the mobile phase has a low viscosity and a high self-diffusion coefficient. This is achieved by using gases such as carbon dioxide, which at critical pressure and critical temperature yield a supercritical fluid with low density. Ammonia and *n*-pentane have also been used. Carbon dioxide at 72.9 atmospheric pressure and 31.3°C has a density of 0.448 g/mL. The mobile phase can be further modified by addition of polar modifiers in small amounts. Thus SFC offers the advantages of both GC and HPLC in that the mobile phase is easy to volatilize, and detectors used in GC can be used here as well; and the mobile-phase selectivity offered by HPLC can still be brought into play by using various additives. The ability to program the density or pressure is unique to SFC in that it allows one to increase the solvating power of the solvent by increasing the density.

Figure 4.8 shows separations for oxazepam by HPLC and subcritical fluid chromatography (SubFC) on the same CSP.[18] A column 15 cm × 4.6 mm with (S)-N-(3,5-dinitrobenzoyl) tyrosine-*n*-butylamide as stationary phase was used. HPLC was performed with 90:10 hexane : ethanol mobile phase at 2mL/min followed by UV detection at 230 nm. SubFC was performed with 92:8 carbon dioxide : ethanol as

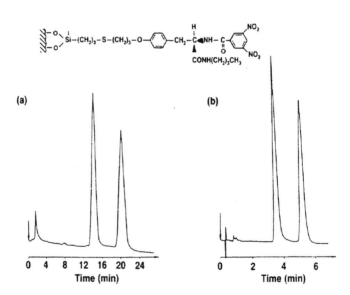

Figure 4.8 Separation of oxazepam by HPLC and subFC. Reproduced with permission from P. Macadiere, M. Caude, R. Rossey, and A. Tambute, *J. Chromatogr. Sci.*, **27**, 583 (1989). Copyright Preston Publications.

mobile phase at a flow rate of 6 mL/min at 200 bar followed by UV detection at 229 nm. Note that separation time is shorter by SubFC, and peak width is narrower than with HPLC. Other examples of SFC separations can be found in chapter 10.

Capillary Electrophoresis

In this technique, a relatively short capillary, attached to the respective reservoirs, is subjected to high voltages around 30 kV (Figure 4.9). The capillary tube has a diameter of 75 μM or less, which allows an easy dissipation of generated heat through the wall. The sample can be drawn inside the capillary tube by a short exposure to high voltage. The zone breadth is proportional to the applied voltage. Capillary electrophoresis has been found to be quite useful for resolving a very large number of compounds, including enantiomers.

The primary advantage of capillary electrophoresis is that it can offer rapid, high resolution of water-soluble components present in small volumes. The separations are based in general on the principles of the electrically driven flow of ions in solution. Selectivity is accomplished by alteration of electrolyte properties such as pH, ionic strength, and electrolyte composition, or by the incorporation of electrolyte additives. Some of the typical additives include organic solvents, surfactants, and complexing agents.

The inherent simplicity of capillary electrophoresis (CE) has prompted many researchers to use it for chiral separations. It offers great potential that is sure to result in many useful applications in the future. Two modes of CE that are commonly used are free solution capillary electrophoresis (FSCE) and micellar electrokinetic chromatography (MEKC)

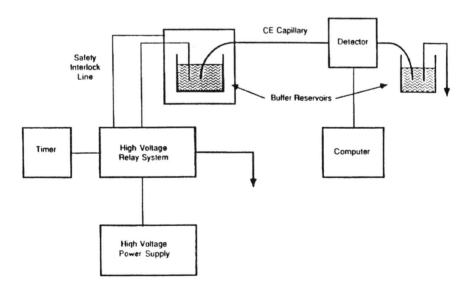

Figure 4.9   Diagrammatic representation of CE.

The use of a carrier electrolyte incorporating specific additives in FSCE is quite useful. The chiral recognition reagents used include cyclodextrins, proteins, carbohydrates, and crown ethers. MEKC is sometimes called MECC (micellar electrokinetic capillary chromatography). Frequently, surfactants such as anionic sodium dodecyl sulfate (SDS) or cationic cetyltrimethyl ammonium bromide (CTAB) are added to the buffer above the critical micelle concentration (typically 50 to 100 mM). Bile salts such as sodium taurodeoxycholate have also been used.

Capillary electrophoresis has been used for direct resolution of terbutaline.[19] (For the structure of terbutaline, see chapter 2.) The chiral environment consists of an aqueous background electrolyte containing β-cyclodextrin or heptakis (2,6-di-O-methyl)-β-cyclodextrin. The effect of the shape and amount of CD added to the background electrolyte on the migration time and the resolution of two enantiomers of terbutaline has been investigated. However, the above electrolyte could not resolve propanolol (**4.3**) isomers. The separation required a background electrolyte urea, β-cyclodextrin, and methanol. Other examples of separations by various modes may also be found in chapter 10.

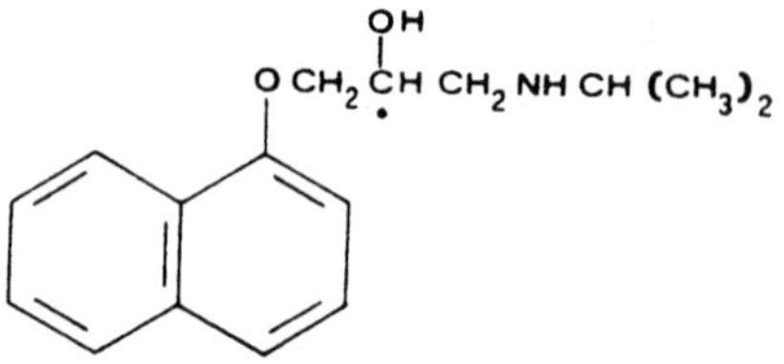

$$\text{O CH}_2\text{CH(OH)CH}_2\text{NH CH(CH}_3)_2$$

4.3

A comparison of modes of separations for chiral compounds used in CE and HPLC is given in Table 4.9. The degree of chiral recognition attainable in chiral CE is often similar to that obtained with HPLC. However, the higher efficiency offered by CE than by HPLC offers a significant advantage. A major advantage of

Table 4.9 Comparison of Modes of Separations CE vs. HPLC

HPLC	CE[a]
Mobile phase additive	FSCE with crown ether
Cyclodextrins	FSCE with cyclodextrins
	MEKC with cyclodextrins
Proteins	FSCE with protein additives
Carbohydrate phase	FSCE with sugar additive
Chiral ligand exchange	FSCE with enantioselective chelation
	MEKC with enantioselective chelation

a. FSCE = Free zone capillary electrophoresis; MEKC = Micellar electrokinetic chromatography; it has been also called micellar electrokinetic capillary chromatography (MECC).

CE is that it is possible to analyze a small sample. Of course, this proves to be a limitation if larger sample size is indicated, as exemplified by preparative separations.

REFERENCES

1. L. Lepri, V. Ceas, and P. G. Deideri, *J. Planar Chromatogr.*, **4**, 338 (1991).
2. R. Bhusan and I. Ali, *J. Chromatogr.*, **392**, 460 (1987).
3. A. Alak and D. W. Armstrong, *Anal. Chem.*, **58**, 584 (1986).
4. M. Remelli, R. Piazzo, and F. Pulidori, *Chromatographia*, **32**, 278 (1991).
5. E. Gil-Av, B. Feibush, and R. Charles, ACS Symposium Series #471. S. Ahuja, Ed. American Chemical Society: Washington, DC, 1991.
6. B. Feibush, A. Figueroa, R. Charles, K. D. Onan, P. Feibush, and B. L. Karger, *J. Am. Chem. Soc.*, **108**, 3310 (1986).
7. S. Allenmark, *Chromatographic Enantioseparation*. Ellis Harwood: New York, 1991.
8. X. Lou, Y. Liu, and L. Zhou, *J. Chromatogr.*, **552**, 153 (1991).
9. T. Kocielski, D. Sybiliska, and J. Jurczak, *J. Chromatogr.*, **280**, 131 (1983).
10. A. Venema and P. J. A. Tolsema, *J. High Res. Chromatogr.*, **12**, 32 (1989).
11. W. A. Konig, R. Krebber, and P. Mischnick, *J. High Res. Chromatogr.*, **12**, 732 (1989).
12. A. Berthod, W. Li, and D. W. Armstrong, *Anal. Chem.*, **64**, 873 (1992).
13. V. Schurig and H. P. Nowotny, *J. Chromatogr.*, **441**, 155 (1988).
14. V. Schurig, M. Jung, D. Schmalzing, M. Schliemer, J. Duvekot, J. C. Buyten, J. A. Peene, and P. Musschee, *J. High Res. Chromatogr.*, **13**, 470 (1990).
15. J & W Applications booklet.
16. D. W. Armstrong, Y. Tang, and J. Zukowski, *Anal. Chem.*, **63**, 2858 (1991).
17. V. Schurig, *J. Chromatogr.*, **666**, 111 (1994).
18. P. Macadiere, M. Caude, R. Rosset, and A. Tambute, *J. Chromatogr. Sci.*, **27**, 583 (1989).
19. S. Fanali, *J. Chromatogr.*, **545**, 437 (1991).

REVIEW

1. List various chromatographic methods that can be used for separation of chiral compounds.
2. Discuss why capillary electrophoresis is not strictly a chromatographic method.
3. Give at least one primary feature of each of the separation methods listed in your response to Problem #1 that would lead you to select that particular method over others.

TUTORIAL

In method selection, consider what is most important for your operation — cost, speed of analysis, ease of operation, availability of equipment, or know-how. Based on these considerations, select a primary method and a backup method. For example, for ease of operation and low cost, TLC is likely to be the method of choice. It can also help do preliminary method evaluations for HPLC. However, it is limited by low efficiency and difficulty in quantifying separated enantiomers. GC, on the other hand, is likely to provide high efficiency; however, it requires that samples are volatile or can be made volatile. HPLC is likely to offer the best choice in terms of resolution, speed, and availability of equipment (most labs these days have at least one HPLC instrument). The limitations are the costs of instrumentation and columns, and HPLC does not offer the highest efficiency. CE and SFC are trying to fill that gap. SFC offers a greater choice of detectors and allows easy collection of separated materials; however, it has limited separation possibilities. CE offers high efficiency, but it can handle only small samples. Precise quantitations require extra efforts, and preparative separations are useful only when very small amounts are needed.

5

Detection Methods

The obvious route for detection of chiral compounds is measurement of optical rotation. However, this measurement value is virtually useless when the sample has more than one enantiomer or is a mixture of chiral compounds. This strongly suggests that the selection of a detection method must relate to the nature of the sample and whether it has been resolved enantiomerically. Listed in Table 5.1 are a number of methods that can be used for determining enantiomeric composition of a sample.

It is often necessary to separate all the chiral compounds present in a sample and to resolve their enantiomers to determine the purity of the chiral compounds. This frequently requires selection of a suitable chromatographic technique first. The choice of detector then relates mainly to the type of chromatographic technique that is being used. Discussed here are the detectors that are commonly used for a given technique.

Detectors for TLC

Most pharmaceutical analysts use TLC as a qualitative or semi-quantitative technique. Detection methods are rather straightforward. These include observation under visible or UV light (short wave or long wave) or spraying the plate with a selected detection reagent. For quantitation, scrape-and-elute methods or electronic scanners are used that allow evaluation of UV, fluorescent, or radiochemical properties of compounds. Visual estimations of small amounts of enantiomers (<1% in some cases) is possible. Radiochemical detection is used mainly to quantify radiolabeled compounds that have been resolved by TLC.

Theoretically, it is possible to use the detectors based on optical rotation for the scrape-and-elute samples. However, sensitivity limitations have to be con-

Table 5.1 Determinations of Enantiomeric Composition

Method	Measurement
Potentiometric	Potential in a cell
Fusion properties	Differential scanning calorimetry
Isotopic dilution	Isotope analysis
Kinetics	Product composition
Diastereotopicity	NMR of diastereomers in achiral solvents
	NMR in chiral solvents
	NMR with chiral shift reagents
Enzyme specificity	Monitor enzyme-catalyzed reaction
Diastereomeric interactions	Chromatography on achiral phase
	TLC
	GC
	HPLC
	Chromatography on CSP
	TLC
	GC
	HPLC
Chiroptical	Optical rotation
	Circular polarization

Source: Reproduced with permission from E. Eliel and S. H. Wilen, *Stereochemistry of Organic Compounds*, Wiley, New York, 1994. Copyright 1994 John Wiley & Sons.

sidered in these cases (see the discussion on optical rotation detectors under HPLC).

Detectors for GC

Gas chromatography is generally used for the separation of diastereomeric mixtures. As already pointed out, the degree of separation of diastereomeric derivatives depends on the derivatization reagent. Alternatively, chiral stationary phases can be used. However, detection is generally performed by two virtually universal detectors: thermal conductivity (TC) or flame ionization (FID). Other specialized detectors based on the unique properties of the analyte are also used when necessary. An added point of interest is that the detectors commonly used for GC can also be used for SFC; thus it is possible to interrelate these separations.

The most commonly used detector is the flame ionization detector for GC, because it offers high sensitivity with ease of handling This detector is destructive to the sample, and its response relates to the carbon content of the sample. A thermal conductivity detector is not quite as sensitive as the FID; however, it is not destructive. Therefore, it allows capture of the eluted solute that can be tested for optical acidity in a separate step. It should be noted that the concentration of eluted solute is likely to be very low, so determination of optical activity is not going to be easy. As a result, HPLC offers a better choice where the optical rotation of the eluted solute can be monitored on line. This is discussed next at some length.

Detectors for HPLC

The conventional detectors used in HPLC can be used for monitoring chiral compounds. Some of these detectors, such as UV or fluorescence, are also used for CE. Basically the same rule applies: the detector should have the required sensitivity to detect the desired sample. Another way of looking at it is that unless we are specifically interested in monitoring optical activity of each of the eluted compounds, the portion of molecule or a molecule as a whole is the best indicator as to which detector is likely to be most useful. We will first look at some of the detectors that are commonly used in HPLC, and then we will discuss the detectors that are best suited for monitoring optical rotation.

Detection in HPLC is complicated by the fact that physical properties of both the mobile phase and the solute can be quite similar. The approaches to detection in HPLC can be classified as follows:[1-4]

- Measurement of a unique sample property not possessed by the mobile phase
- Differential measurement of a general property of both the sample and the mobile phase
- Detection based on a unique sample property after removal of the mobile phase

A number of detectors have been developed based on one of these approaches. An ideal detector should have the following characteristics:[4]

- Responds to all solutes or at least has predictable specificity
- Exhibits high sensitivity with a constant predictable response
- Shows a wide range of linearity
- Is unaffected by changes in temperature or flow rate
- Responds independently of the mobile phase
- Is nondestructive of solute
- Responds quickly
- Does not contribute to band broadening
- Provides qualitative information on the eluted peak

Unfortunately, no HPLC detector meets all of these requirements; however, the information provided here can serve as a useful index in selecting an appropriate detector for a given purpose. Some of the detectors that have been used in HPLC include UV, fluorescence, electrochemical, conductivity, refractive index, mass spectrometer, and miscellaneous detectors.

It is important to remember that separation in HPLC is a dynamic process and requires a detector that can work on-line and produce an immediate record of column eluents. This requires that the detector have good sensitivity and small volume to avoid additional band broadening. Furthermore, it should have rapid response that can monitor rapidly changing eluent concentrations. To enable the reader to make a good selection, the specifications for some of the commonly used commercial detectors are given in Table 5.2.

Table 5.2 Specifications of Commercial HPLC Detectors

Property	UV/Visible	Fluorescence	Electochemical	Conductivity	Refractive Index
Mass (ng)	0.1–1	10^{-2}–10^{-3}	0.01–1	0.5–1	100–1000
Conc. (g/mL)	10^{-8}	10^{-11}	10^{-10}	10^{-8}	10^{-7}
Linear range	10^5	10^3	10^6	10^4	10^4

Of the various detectors listed in Table 5.2, only UV/visible and fluorescence detectors allow gradient performance. That's why these detectors are so frequently used in HPLC. Of the two, UV is by far the more popular because of the ease of operation and greater applications to a variety of samples. The comparability of response of a typical chiral compound to a UV versus an optical detector (see Specific Detectors for Chiral Compounds) is shown in Figure 5.1.

A mass spectrometer is used as a detector when it is desirable to obtain structural information on an impurity. It does not offer any significant advantage for detecting or quantitating chiral compounds aside from the fact that if such an impurity is seen, it is possible to deduce from the fragmentation pattern its relationship to the parent structure. There are significant difficulties in using this detector because buffers used for HPLC present problems. If buffers must be used, it is possible to use only those that can be volatilized. The flow rates employed for HPLC are generally fairly high, to allow direct introduction of the eluted solute into MS for determination by HPLC/MS. Stream splitting can be employed for introduction into MS, but this can significantly decrease detectabilities.

A number of other detectors, such as atomic absorption spectrometer, electron capture, evaporative light scattering, and low-angle laser light scattering, have been employed for specialized situations. However, it should be emphasized that these detectors are not specific for chiral compounds.

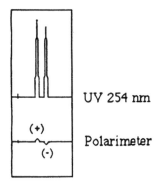

Figure 5.1 Typical responses of chiral and UV detectors.

Specific Detectors for Chiral Compounds

The detectors that employ optical rotation as a means of detection have been called chiroptical detectors. These detectors are specially suited for the detection of resolved enantiomers.

Let's review what is involved with these detection methods. The properties of chiral substances arising from their nondestructive interaction with anisotropic radiation (polarized light) are called chiroptical properties and can allow one to differentiate between the two enantiomers of chiral compounds. The use of this term was introduced by Thomson in 1884.[1] It encompasses the classical spectroscopic qualitative and quantitative manifestations of chirality in terms of optical activity and optical rotatory dispersion (ORD).

Optical activity or optical rotation results from the refraction of right and left circularly polarized light to different extents by chiral molecules. The source of rotation is birefringence, the unequal slowing down of right and left circularly polarized light as the light passes through the sample. By definition, this is also the case with ORD, albeit at a given wavelength. ORD relates to the change of optical rotation with the wavelength of light.

Widespread use of another chiroptical technique, circular dichroism (CD) is more recent. CD relates to the difference in absorption of right and left circularly polarized light. The accessible chromophore of the molecule plays a role in this technique as well as with the emission of circularly polarized light (CPE); however, it has no effect on optical rotation or ORD.

Vibrational CD and its counterpart in Raman spectroscopy are relatively new chiroptical techniques being developed now, along with CPE.

Optical rotation can be seen as the result of interaction of light with the sample that produces electric and magnetic moment changes that are not at right angles to one another. A rotation to the right (dextrorotation) results when these effects are parallel, and rotation to left (levorotation) occurs when they are antiparallel. This behavior results from the chiral features of the molecular architecture that impose helicity on the motion of electrons. When the sample contains an equal number of dextro and levo molecules of a given structure, as is the case with racemates, this leads to zero rotation. Because each molecule individually rotates the plane of polarization from that of the incident radiation, the net rotation is zero—the number of molecules rotating in one direction equals those rotating in the opposite direction. Intermediate situations exist for mixtures of chiral molecules containing an excess of one of the enantiomers.

A simple device for measuring optical rotation is a polarimeter (see Figure 5.2). The observed optical rotation can be measured with a polarimeter and is proportional to the length of the cell (l) and the concentration (c) of the solution.

Specific Rotation

The specific rotation [α] of a liquid is defined as the angular rotation in degrees through which the plane of polarization of polarized monochromatic light is ro-

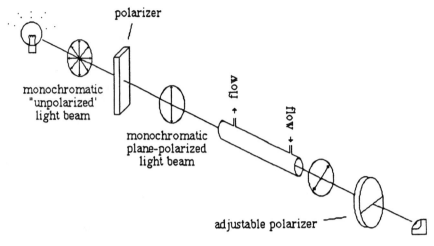

Figure 5.2 Diagrammatic representation of a polarimetric detector.

tated by passage through 1 decimeter (100 mm) of the liquid, calculated on the basis of specific gravity of 1. In the case of solutions of an optically active substance, the specific rotation is calculated on the basis of a concentration of 1 g of solute in 1 mL of solution. Since the specific gravity of liquids, as well as the optical activity of many substances, is influenced by temperature, it is necessary to indicate the temperature at which rotation and specific gravity are determined. For most purposes, the specific rotation is reported at the wavelength of sodium D-line (589 nm) at 25°C by the equation reproduced here from chapter 3.

$$[\alpha]_D^t = \frac{100\,a}{lpd}$$

where a = observed rotation in degrees of a solution at temperature t using sodium light, l = length of tube in decimeters, p = concentration of solution expressed as the number of grams of active substance in 100 mL of solution, and d = density of solution at temperature t.

Obviously $[\alpha]$ values can be determined at different temperatures and at different wavelengths; it is important that these values be clearly indicated.

Advantages of Optical Rotation Detectors

The methods that use an optical rotation detector offer the following advantages:

1. They can be used to determine whether the resolved chemical species is optically active. For example, if a stereoselective synthesis is being attempted from a racemate, optical activity can inform of the progress. Alternatively, chiroptical detection can determine whether or not an

enantiomer has completely racemized as a result of the procedure with time.
2. When used as a polarimeter, chiroptical detection can provide a check for a product's optical purity. A reference standard is generally necessary for this purpose.
3. An optical rotation detector can help identify/validate a specific stereoisomer. Optical rotation detectors can also provide an additional clue to the identity of a particular optical isomer, for example, if it is necessary to determine which of an enantiomer pair is present as an impurity.
4. Optical rotation detectors are useful for quantitative measurement of "poorly resolved" stereoisomers. It is frequently possible to obtain useful quantitative information from fused chromatographic peaks with an ORD.
5. Chiroptical detection may be especially useful for determination of stereoisomers with weak chromophores.
6. In a process development sample, an optical rotation detector in conjunction with another one, such as a UV detector, will indicate the optically active chromatographic peaks of a sample.
7. These detectors can be useful for researching methods to obtain a desired elution order of optical isomers in chromatography.

These detectors work well for many compounds with large specific rotations; however, many compounds produce a poorly detectable signal with typical sample injections, especially at the high wavelength used in these detectors.

The standard option of increasing path length is not generally available with the polarimetric detectors for HPLC. So the only option is to increase the amount of solute in the injected sample — an option that does not allow analysis of low-concentration samples or samples that are available in minute amounts.

Sample Detectability

A simple calculation that can give some insight into whether or not a particular sample will be detectable by the optical rotation detector can be done as follows:[5] If injection volume (μl) × concentration (%) × specific rotation (degrees) is greater than 100, there is a very good possibility that the detector will be sufficiently responsive.

The specifications of a typical polarimetric detector (Chiralyser) are as follow:

Operation principle: Faraday compensation

Measuring range: 32×10^{-4} degrees minimum 4096×10^{-4} degrees maximum in 8 ranges

Resolution: 12 bit/full scale

Noise: 2.5×10^{-5} degrees

Stability: 5×10^{-5} degrees

Limit of detection: 0.5 μg on column for $\alpha = 90$ degrees

Average: 2–250 single cycles/measurement

Autozero: zero point setting by key contact

Light source: halogen lamp

Optical system: polychromatic system with coated optics, center wavelength of 546 nm

Flow cell: glass-coated stainless steel, optical length 200 mm

Interface: chart recorder ± 0.5 V

These specifications are provided here to enable readers to select a polarimetric detector. Limit of detection along with low noise level and stability are some of the obvious considerations in the selection process.

ORD Detectors

ORD detectors scan a range of wavelengths. An ORD curve is obtained when the optical activity of a chiral compound is measured and plotted as a function of wavelength. When no chromophore is present in the sample, the optical rotation will decrease continuously with increasing wavelength. If, on the other hand, the spectral range investigated covers an absorption band of the compound, that band will give rise to a Cotton effect—the curve will show one or more peaks or troughs.

Here are a few important points to remember when selecting a flow cell to be used for ORD:

- A flow cell with a long path length is needed to obtain a good signal.
- The cell volume must be kept to a minimum to avoid peak dispersion.
- Very narrow flow cells present significant engineering problems.
- Internal reflections degrade beam polarization.

It is good to remember during evaluations of an ORD detector for HPLC that a minimum sensitivity of 0.1 millidegree is desirable.

Circular Dichroism Detectors

CD offers some inherent advantages over ORD as a chiroptical detection method because differential absorbance of right and left circularly polarized light is measured at a number of wavelengths. It is possible to scan all wavelengths, so signal strength is much less of a problem as compared to ORD detectors. It may be recalled that ORD detectors utilize a fixed wavelength, which is frequently not the wavelength at which maximum absorbance is observed. CD spectra provide valuable information that can be useful for assignment of absolute configuration for the unknown compounds. They allow measurement of differential dichroic absorption and molecular ellipticity as a function of wavelength. Frequently, the Cotton effect is better evaluated from a CD curve. Dedicated CD HPLC detectors are under development to make them price-competitive with polarimetric detectors. HPLC flow cells are also available for the conventional CD instruments. These detectors are quite expensive (around $50,000) and are likely to remain that way for the foreseeable future.

The relationship of ORD and CD spectra to the UV spectrum is shown in Figure 5.3. This figure shows the sign and magnitude of the Cotton effect. The compound studied has two chromophores with absorption maxima at 217 and 293

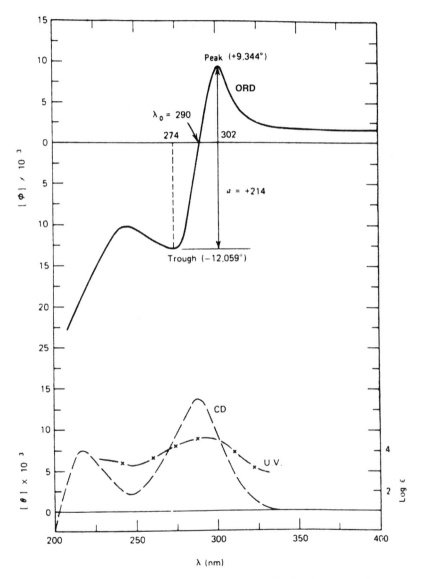

Figure 5.3 Comparison of ORD, CD, and UV spectra of a chiral compound. Reproduced with permission from P. Crabbe and A. C. Parker, in *Physical Methods of Chemistry*, A. Weissberger and R. W. Rossiter, Eds., Wiley, New York, 1972, Part 3, p. 183. Copyright 1972 John Wiley & Sons.

nm. If we consider only the latter band, it is readily seen that this shows a positive Cotton effect ($\alpha = 214$) and that the λ_0 value (290 nm) corresponds to the CD-maximum and UV-absorption maximum. If the optical antipode of this compound had been investigated, the ORD and CD spectra would have been completely inverted in sign along the x-axis.

REFERENCES

1. As cited in E. Eliel and S. H. Wilen, *Stereochemistry of Organic Compounds*. Wiley: New York, 1994.
2. S. Ahuja, *Selectivity and Detectability Optimizations in HPLC*. Wiley: New York, 1989.
3. E. S. Yeung, *Detectors for Liquid Chromatography*. Wiley: New York, 1986.
4. L. R. Snyder, J. L. Glajch, and J. J. Kirkland, *Practical HPLC Method Development*. Wiley: New York, 1988.
5. Bulletin on Chiralyser, JM Science, Grand Island, NY.

REVIEW

1. What is a chiroptical detector?
2. Define optical rotation.
3. Provide an equation for specific rotation.
4. Is it always necessary to use chiroptical detector for the analysis of chiral compounds?

TUTORIAL

The primary advantage that chromatographic methods offer in separations of chiral compounds is resolution — that is, the two components of interest are sufficiently separated from each other and other components, if any, in the sample to allow reliable quantitations. Once a satisfactory separation has been achieved, it is necessary to assure that the separated components are indeed what we believe them to be. This is best accomplished by comparison with known standards. Combination with mass spectrometry might also help deal with identification questions. Once assurance has been obtained on the identification of the material, then it is not necessary to use chiroptical detectors. Any detector that provides the greatest detectability can be used. The optical rotation detectors offer some unique advantages (reproduced here from the section on sample detectability):

1. They can be used to determine whether the resolved chemical species is optically active. For example, if a stereoselective synthesis is being attempted from a racemate, optical activity can inform of the progress. Alternatively, chiroptical detection can determine whether or not an enantiomer has completely racemized as a result of procedure with time.
2. When used as a polarimeter, chiroptical detection can provide a check for a product's optical purity. A reference standard is generally necessary for this purpose.
3. An optical rotation detector can help identify/validate a specific stereoisomer. Optical rotation detectors can also provide an additional clue to the identity of a particular optical isomer, for example, if it is necessary to determine which of an enantiomer pair is present as an impurity.

4. Optical rotation detectors are useful for quantitative measurement of "poorly resolved" stereoisomers. It is frequently possible to obtain useful quantitative information from fused chromatographic peaks with an ORD.
5. Chiroptical detection may be especially useful for determination of stereoisomers with weak chromophores.
6. In a process development sample, an optical rotation detector in conjunction with another one, such as a UV detector, will indicate the optically active chromatographic peaks of a sample.
7. These detectors can be useful for researching methods to obtain a desired elution order of optical isomers in chromatography.

A good rule to remember is that if injection volume (μl) × concentration (%) × specific rotation (degrees) is greater than 100, there is a very good possibility that the detector will be sufficiently responsive.

6

Desirable Features of Chiral Stationary Phases

There are a number of advantages to carrying out separations on a chiral stationary phase (CSP), not the least of which is that derivatization to form diastereomeric derivatives is generally not required. In other words, the use of a chiral stationary phase allows direct separations with minimal sample preparation, and selective interactions can be encouraged by appropriate selection of the stationary phase. Furthermore, preparative separations are more easily achieved.

Desirable Features

The following list of advantageous features of a CSP may prove to be quite useful in the selection of a suitable CSP.[1] A short discussion on the listed items follows.

- Baseline resolution of enantiomers
- High efficiency
- Short analysis time
- Column stability (time, temperature, solvent)
- Mobile-phase compatibility
- Ability to invert elution order
- High capacity

Baseline Resolution

The following equation is most commonly used by the practicing chromatographer to calculate resolution (Rs):

$$Rs = \frac{t_2 - t_1}{0.5\,(t_{w1} + t_{w2})} \qquad (6.1)$$

where t_1 = retention time of the first peak, t_2 = retention time of the second peak, t_{w1} = peak width at baseline for the first peak, and t_{w2} = peak width at baseline for the second peak

An Rs value of 1.5 yields baseline resolution and allows convenient determination of enantiomeric excess (e.e.) values. If the Rs value is <1.5, the peaks are likely to overlap and determination of the enantiopurity value is made more difficult. It may be recalled that enantiopurity is usually reported in terms of enantiomeric excess (% e.e.).

$$\% \text{ e.e.} = \frac{\text{area of major peak} - \text{area of minor peak}}{\text{area of major peak} + \text{area of minor peak}} \times 100 \qquad (6.2)$$

Resolution values that are much larger than 1.5 are very useful for preparative separations because the contamination from the related component can be eliminated. The calculations for e.e. values can be complicated by different peak shapes (it is well recognized that peak shapes can be quite different for various separations).

When the peak width at half height is used in Equation 6.1 instead of the peak width at the base of the peak, the divisor 0.5 is replaced by the multiplier 1.198.

Complex equations that derive results from first principles utilize efficiency (N), separation factor (α), and capacity factor (k') in the calculations may be found in some of the standard texts on chromatography. A simplified equation is given here.

$$\text{Resolution} = Rs = 0.25 N^{1/2} \left[\frac{\alpha - 1}{\alpha} \right] \left[\frac{k'}{1 + k'} \right] \qquad (6.3)$$

where N = number of theoretical plates (see Equation 6.4), α = separation factor (k'_2/k'_1), and k' = peak capacity.

An average value for peak capacity (k') of the two peaks can be used in the above equation; however, many chromatographers prefer to use k'_2 (retention time of peak 2 − retention time of unretained peak divided by retention time of peak 2) because it is likely to be equal to or slightly greater than the average and thus gives a better measure of N.

Efficiency

Tailing or distorted peaks make separations very difficult, especially for the purposes of preparative separations. Tailing is related to lack of homogeneity of the column, overloading, unusual adsorption sites, and a host of other factors. The tailing problem can be circumvented by designing more efficient columns. The term *efficiency* (N) refers to the number of theoretical plates available in a column. Highly efficient columns with theoretical plates more than 100,000 per meter are available in chromatography. However, for chiral separations, the columns with this high N value are yet to be designed. The columns commonly used for chiral separations

Table 6.1 Minimum Number of Plates Required to Obtain Rs of 1.0

k'_1	$\alpha = 2.00$	$\alpha = 1.10$	$\alpha = 1.01$
0.1	1936	193,600	1,936,000
1	64	6,400	640,000
10	19.4	1,936	193,600

Source: Reproduced from S. Ahuja, W. Pirkle, and C. Welch, *Chiral Separations by Chromatography*, American Chemical Society Short Course, 1994. Copyright 1994 American Chemical Society.

provide adequate efficiency to provide peaks with narrow width at the baseline that allow separations to be carried out in a short time.

The total number of theoretical plates can be calculated by the following commonly used equation:

$$N = 16\,(t/w)^2 \tag{6.4}$$

where t = retention time of the peak of interest and w = peak width at the baseline.

When the peak width at half height is employed in the above equation, the multiplier 16 is replaced by 5.54.

To get a better measure of Rs, it is necessary to use Equation 6.3, where α and k' are also included. Based on this equation, the minimum N required can be calculated for any value of Rs. The minimum number of required theoretical plates for a nominal Rs value of 1.0 is given in Table 6.1. The number of theoretical plates for desired higher values of Rs can be calculated from Equation 6.3, recognizing that Rs is proportional to the square root of N. As the table indicates, for low k' and low α values, the N values have to be very large. For example, almost 2 million theoretical plates are required to bring about a separation with $k' = 0.1$ and $\alpha = 1.01$.

Figure 6.1 shows the relative resolution power of various chromatographic techniques. It is apparent that for a given chiral selector, more racemates can be resolved using techniques that afford greater chromatographic efficiency.

Rapid Analysis

The following factors help to achieve rapid analysis time: high column efficiency, high enantioselectivity, and a wide range of mobile-phase usage. All of these factors are likely to lead to a narrower peak width and/or shorter analysis time.

Column Stability

A stable CSP column has the following characteristics:

- Stable backbone — the "inert" material that is called chromatographic support

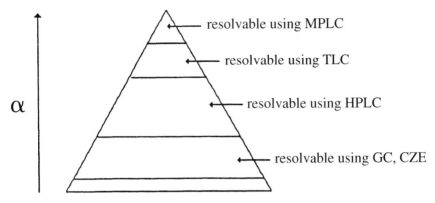

Figure 6.1 Resolution with various chromatographic techniques. Reproduced from S. Ahuja, W. Pirkle, and C. Welch, *Chiral Separations by Chromatography*, American Chemical Society Short Course, 1994. Copyright 1994 American Chemical Society.

- Covalent bonding of selector (not adsorbed) to support
- No labile linkage or groups such as esters or carbamates
- Absence of stereogenic centers that are easily racemized
- No superfluous adsorption sites

Mobile-Phase Compatibility

Ideally a CSP column should work in both the normal and the reversed-phase modes. It should also tolerate a wide variety of solvents such as water, hexane, methanol, dichloromethane, and so on without loss of performance.

Inversion of Elution Order

The ability to invert the elution order of the enantiomers by changing to an identical CSP of the opposite absolute configuration is very useful for determination of trace components, especially if they elute after the main component (Figure 6.2). This approach is very helpful in the preparative separations for minor components.

High Capacity

Low-capacity CSPs are easily overloaded, which makes them unsuitable for preparative separations. Since only low amounts of sample can be injected to obtain a useful separation, detection can also be a problem. It is desirable to have a high density of binding sites on the CSP because high capacity allows high loading. This permits injection of a larger amount of sample without changes in column performance.

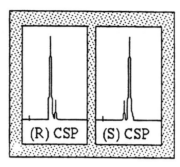

Figure 6.2 A typical representation of inversion of elution order. Reproduced from S. Ahuja, W. Pirkle, and C. Welch, *Chiral Separations by Chromatography*, American Chemical Society Short Course, 1994. Copyright 1994 American Chemical Society.

Types of CSPs

The CSPs commonly used for enantiomeric separations by HPLC can be classified as follows: polysaccharides, cyclodextrins, proteins, ligand exchangers (LE), chiral polymer, crown ethers, and brush-type.

Polysaccharides and Derivatives

The chirality of natural products, especially carbohydrates, makes them desirable candidates for optical resolution by liquid chromatography. In 1938, a partial resolution of camphor derivative was obtained on a column packed with lactose.[2] The resolving power of cellulose, a polysaccharide, was discovered by the observation that a racemic acid can occasionally produce two spots in paper chromatography. The three-point interaction theory of chiral compounds was put forth by C. E. Dalgliesh in 1952, on the basis of results from paper chromatography of racemic amino acids.[3] This led to utilization of cellulose in TLC, and eventually in HPLC.

Cellulose is a linear polysaccharide (see Figure 4.3). It forms very long chains containing 1,500 (+)-D-glucose units per molecule with a molecular weight range of 250,000 to 1 million. These long molecules are arranged in parallel bundles in a cellulose fiber and are held together by numerous hydrogen bonds between various hydroxyl groups. The structure of a (+)-D-glucose unit shows that it contains five chiral centers and three hydroxyl groups, and all three ring substituents are equatorial. Treatment with dilute alkali can cause cellulose to lose its enantioselective properties due to the transformation of native, metastable form into a rearranged and stable amorphous form.

Starch is also a polysaccharide made up of (+)-D-glucose units, but its structure is more complex than that of cellulose (Figure 6.3). It contains approximately 20% amylose and 80% amylopectin. Amylose is a linear polymer; however, amylopectin is branched and constitutes the insoluble fraction of starch. Starch is useful for separation of polar compounds.

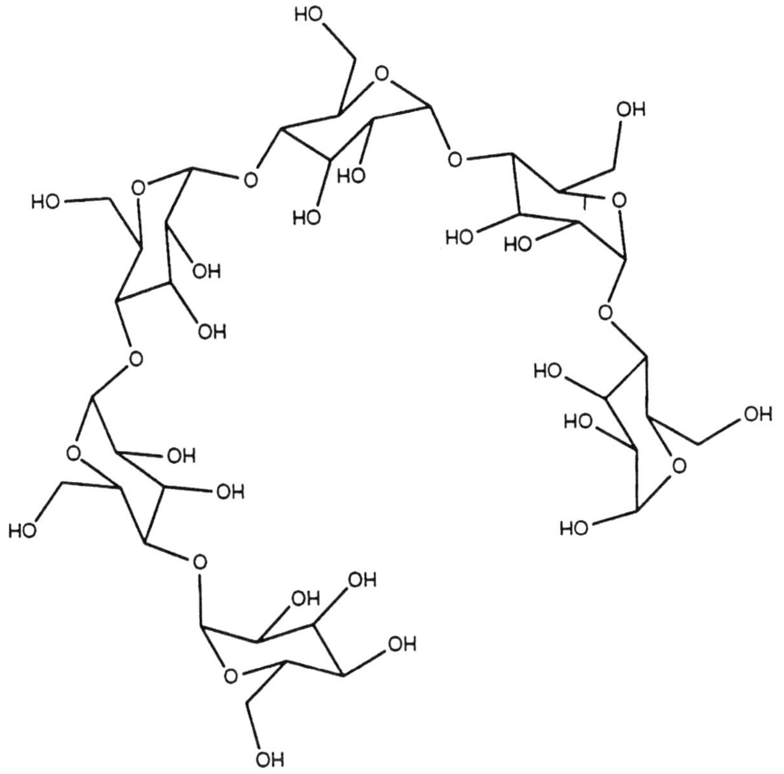

Figure 6.3 Starch.

Partially acetylated cellulose was found to provide resolution of optical isomers in column chromatography in the late 1960s. In the 1970s, researchers investigated heterogeneous acetylation of native (microcrystalline) cellulose and found that cellulose triacetate could be prepared with almost complete preservation of microcrystallinity and excellent resolving properties[4] — a discovery that resulted in the use of chemically modified polysaccharides as CSPs. These latter experiments led to the conclusion that microcrystallinity is essential for the enantioselective properties of these materials because the optical resolvability is lost on dissolution and reprecipitation. The metastable nature of these materials is evident from the results of these experiments, as the change has been found to be irreversible. Furthermore, these investigations suggest that retention occurs by inclusion of the solute into the molecular cavity — thus giving rise to the term *inclusion chromatography*. This proposal is supported by the difference in behavior between benzene and mesitylene (1,3,5-trimethylbenzene). Benzene is retained much more strongly because of better permeation into cavities.

Currently used modified polysaccharide CSPs consist of a chemically modified polysaccharide (starch or cellulose) adsorbed on silica chromatographic sup-

Table 6.2 Resolution on Cellulose Triacetate Columns (250 × 4.6 mm; eluent: ethanol)

Compound	Sorbent	k'_1	k'_2	α
[structure: dibenzodiazepine]	I	2.61 (−)	5.36 (+)	2.05
	II	0.59 (+)	0.91 (−)	1.53
[structure: 2,3-diphenyloxirane]	I	7.82 (+)	11.3 (−)	1.45
	II	0.94 (−)	1.23 (+)	1.31
Ph−CH−CONH$_2$ | OH	I	2.08	3.08	1.48

a. I = microcrystalline cellulose triacetate; II = reprecipitated cellulose triacetate.
Source: Data from G. Hess and R. Hazel, *Chromatographia*, 6, 277 (1973).

port. The k' values of these materials are generally lower than those of the parent material; however, column efficiency is much higher and compensates for the lower α values. These investigations show that the microcrystallinity of cellulose triacetate is not an absolute requirement for efficient chiral resolution. As a result, new derivatives are continually being introduced.

Table 6.3 A Variety of Chiral HPLC Columns

Name	Type of Adsorbent
Chiralcel OJ	Cellulose ester derivative
Chiralcel OB	Cellulose ester derivative
Chiralcel OB-H	Cellulose ester derivative
Chiralcel OA	Cellulose ester derivative
Chiralcel OK	Cellulose ester derivative
Chiralcel CA-1	Cellulose ester derivative
Chiralcel OD	Cellulose carbamate derivative
Chiralcel OD-H	Cellulose carbamate derivative
Chiralcel OC	Cellulose carbamate derivative
Chiralcel OG	Cellulose carbamate derivative
Chiralcel OF	Cellulose carbamate derivative
Chiralpak AD	Amylose derivative
Chiralpak AS	Amylose derivative
Chiralcel OD-R	10 μm, reversed-phase type of chiralcel OD
Chiralcel OJ-R	5 μm, reversed-phase type of chiralcel OJ
Chiralpak OT (+)	Polymethacrylate
Chiralpak OP (+)	Polymethacrylate
Chiralpak WH	Ligand exchange
Chiralpak WM	Ligand exchange
Chiralpak WE	Ligand exchange
Chiralpak MA (+)	Ligand exchange (coating type)
Crownpak CR	Crown ether

Four types of derivatives can be easily prepared by modification of the free hydroxyl groups: carbamates, esters, ethers, and nitrates. Table 6.2 shows resolution of a variety of different compounds with these materials.[5] Some of the useful cellulose derivatives are summarized below.

- Triacetate: especially effective for substrate with phosphorus atom as a stereogenic center
- Tribenzoate: useful for racemates with a carbonyl group near a stereogenic center
- Trisphenylcarbamate: suitable for polar racemates and sensitive to molecular geometry
- Tribenzyl ether: effective with protic solvents as mobile phases
- Tricinnamate: useful for aromatic racemates and barbiturates (yields high retention times)

A variety of CSPs are available, as shown in Table 6.3. The structures of some of the commonly used CSPs from this group are shown in Figure 6.4. The advantages and disadvantages of these CSPs are as follows:

Advantages

    A wide variety of enantiomers can be resolved.

    They offer reasonable efficiency.

    They provide good capacity.

Disadvantages

    The CSPs provide poor stability.

    They have poor solvent compatibility.

    It is not possible to invert elution order.

    They are sensitive to high pressure and high flow rates.

    The structure-resolution properties are difficult to predict.

Cyclodextrins

Dextrins are formed when starch is subjected to degradation by *Bacillus macerans*. These compounds are composed of β-1,4-D-glucosides that have been cyclized to a ring of 6 to 12 units. The molecules with 6, 7, and 8 units have been called α-, β-, and γ-cyclodextrins, respectively. Cyclodextrins (CD) can form inclusion complexes with various compounds of appropriate size. In HPLC, β-cyclodextrin (Figure 4.1) is most commonly used because it is easily available and has offered a great deal of promise. The diameter of β-CD ring is 8 Å, and its volume is approximately 350 Å.[3]

It appears that CD columns operate in the reversed-phase mode, since the formation of inclusion complexes with CDs in an aqueous system is based mainly on hy-

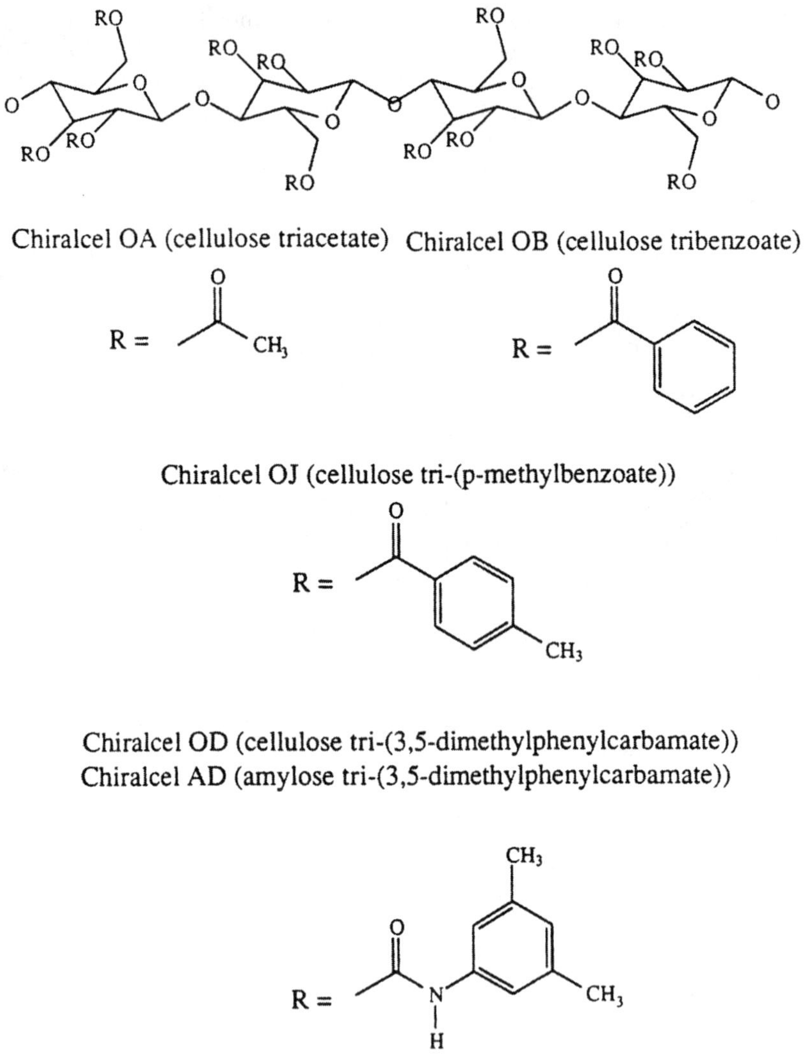

Figure 6.4 Structures of Daicel CSPs.

drophobic interactions. Therefore, the mobile-phase systems generally used in RPLC can be used. The mobile phases usually contain methanol or acetonitrile with water. Buffers can be used to control pH and influence the retention of ionizable solutes. The effect of the mobile phase on enantioselectivity is quite significant. Both the k' and α values tend to decrease with increasing content of organic modifier. Retention is also significantly affected by temperature, approaching zero around 60°C or so.

Cyclodextrins are used as stationary phases for both GC and HPLC. They are also used as mobile-phase additives. The advantages and disadvantages of cyclodextrins as CSPs are given here.

Advantages

> They offer high capacity.
>
> These CSPs are quite stable.
>
> A wide range of solvents can be used.

Disadvantages

> The applications are limited to compounds that can enter the cyclodextrin cavity. Many exceptions have been found, especially with modified cyclodextrins, where other interactions can also occur.
>
> Small changes in analyte structure can often lead to unpredictable effects upon resolution.
>
> It is not possible to invert the elution order.
>
> They often offer poor efficiency.

Proteins

It has been known for some time that binding to proteins involves multiple equilibria because proteins have a number of binding sites and some of those sites are likely to have a different affinity for the ligand. A protein selector is immobilized on a chromatographic support. The transport proteins show a broad generality in chiral recognition of a great variety of enantiomers. Some of the proteins that have been used include albumins (BSA, HSA), α-1 acid glycoprotein, and ovalbumin.

Other proteins, including antibodies, have been used as CSPs. There are ongoing developments in choice, stabilization, and immobilization of proteins. The advantages and disadvantages of these CSPs are listed below.

Advantages

> They are often useful for a broad range of pharmaceutical compounds that are resolvable.
>
> These CSPs provide reasonably high efficiency.

Disadvantages

> They provide very low capacity and therefore are not suitable for preparative separations of significant amounts of the material.
>
> Solvent compatibility is generally poor.
>
> The elution order cannot be inverted.
>
> They offer poor mechanical stability (improvements are being made in this area).
>
> Only low flow rates can be used, engendering long analysis times.

Ligand Exchangers

These CSPs are derived from amino acids that are capable of resolving enantiomers of underivatized amino acids through formation of metal ions, usually cupric ion, chelate (see Figure 6.5). A number of commercial suppliers provide these materials. They offer the following advantages and disadvantages:

Advantages

Copper chelate serves as a chromophore.

No derivatization is required; separation of underivatized amino acids is possible.

They offer good efficiency.

Disadvantages

Their usefulness is limited to compounds that can form bidentate ligands — amino acids or hydroxy acids, for example.

They require a mostly aqueous mobile phase.

Preparative separations are not easily carried out because of the presence of cupric ions in the eluent.

Chiral Polymers

These can be broadly classified into two groups: polyacrylamides and polymethacrylamides.

Polyacrylamide and polymethacrylamide derivatives have been used where the chiral substituents originate from an optically active amine or amino acid component. Polymer particles of the desired mean diameter and acceptable size and homogeneity can be obtained by utilizing suspension polymerization techniques.[6]

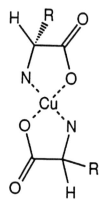

Figure 6.5 A typical representation of a chelate.

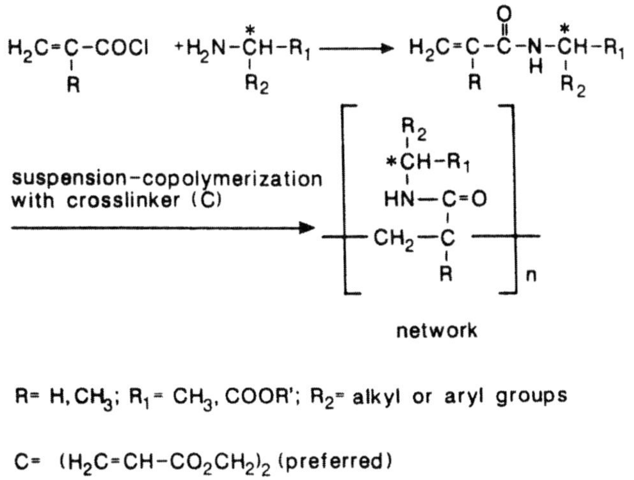

$$H_2C=C-COCl + H_2N-\overset{*}{C}H-R_1 \longrightarrow H_2C=C-\overset{O}{\overset{\|}{C}}-\underset{H}{N}-\overset{*}{C}H-R_1$$
$$RR_2RR_2$$

suspension-copolymerization with crosslinker (C) →

$$\left[\begin{array}{c} R_2 \\ | \\ *CH-R_1 \\ | \\ HN-C=O \\ | \\ -CH_2-C- \\ | \\ R \end{array}\right]_n$$

network

R= H, CH$_3$; R$_1$= CH$_3$, COOR'; R$_2$= alkyl or aryl groups

C= (H$_2$C=CH-CO$_2$CH$_2$)$_2$ (preferred)

Figure 6.6 Synthetic route of polyacrylamide and polymethacrylamide. Reproduced with permission from T. Shibata, I. Okamoto, and K. Ishi, *J. Liq. Chromatogr.*, 9, 313 (1986).

Free-radical initiation is used, and porosity of the gel is regulated by the relative amount of cross-linking agent added. Since these particles swell in organic solvents, these materials can be used only in low-pressure LC systems.

The resolution ability of these polymers is dependent on a number of factors such as substituents on the polyacrylamide backbone, the degree of crossing, the nature of cross-linking, and the mobile-phase composition. A preferred route to preparation of these polymers is given in Figure 6.6. These polymers have been found particularly useful for polar compounds with functional groups that can form hydrogen bonding.

Okamoto et al. prepared a vinyl polymer with chirality based on its helicity.[7,8] Optically active polymethacrylate (triphenylmethyl methacrylate) is prepared by asymmetric anionic polymerization of triphenylmethyl methacrylate under the influence of a chiral initiator in toluene at low temperatures (Figure 6.7). The polymer is insoluble in most common organic solvents at a degree of polymerization greater than 70. The advantages and disadvantages of these CSPs are given here. As a result of these disadvantages, their usefulness is limited.

Advantage

They are useful for separating racemates containing aromatic "propeller-type" chirality.

Disadvantages

They exhibit poor efficiency.

Figure 6.7 Synthesis of poly (triphenylmethyl methacrylate). Reproduced with permission from T. Shibata, I. Okamoto, and K. Ishi, *J. Liq. Chromatogr.*, 9, 313 (1986).

They exhibit instability in methanol, a commonly used mobile-phase solvent.

They need special storage (hexane) and operating conditions (5°C).

Crown Ethers

Macrocyclic polyethers are known as crown ethers because molecular models of them resemble the shape of a crown (Figure 6.8). They are prepared by introduction of an enantio-enriched spacer unit (most commonly a binaphthol unit). The advantages and disadvantages are as follows:

Advantages

They are useful for separations of underivatized primary amines and amino acids.

These can be used for ion chromatography.

Disadvantages

They offer poor efficiency.

They are unsuitable for preparative separations.

The eluent perchloric acid has a pH value around 1, which is quite harsh for a number of compounds.

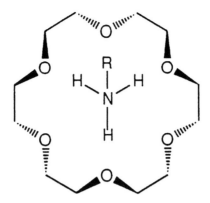

Figure 6.8 Formation of a complex with a crown ether.

Imprinted Polymers

The preparation of these polymers can be shown diagrammatically, as in Figure 6.9. The process can be assumed to occur in three steps:

1. A complex is formed between the chiral compound used as template and a polymerizable monomer.
2. The polymerization process with cross-linking produces a rigid matrix.
3. The template is removed by washing or by a simple reaction such as hydrolysis.

A few examples of template molecules and their resolution are given in Table 6.4.[9] The advantages and disadvantages of these CSPs are given here.

Advantages

  They offer potential for high enantioselectivity

  They can be custom-tailored for particular compounds.

  They are stable.

  Either enantiomeric form can be made.

  Solvent compatibility is not a problem.

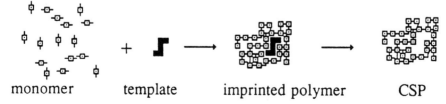

Figure 6.9 Diagrammatic representation of imprinted polymer preparation. S. Ahuja, W. Pirkle, and C. Welch, *Chiral Separations by Chromatography*, American Chemical Society Short Course, 1994. Copyright 1994 American Chemical Society.

Table 6.4 Resolution on Imprinted Chiral Polymers

Polymer and print molecule	Particle diameter (μm)	Column temp. (°C)	k'_2 (configuration)	α	Rs
A, L-PheNHPh	10–30	80–85	4.3 (L)	3.2	1.2
C, L-*p*-H$_2$NPheOEt	10–30	80–85	2.7 (L)	1.8	0.8
E, L-PheNHPh	32–45	80–85	5.1 (L)	4.9	1.0
G, L-TrpOEt	45–65	80–85	1.2 (L)	1.8	0.5
H, L-PheNHPh	32–45	23	1.2 (L)	3.7	0.9
I, D-*p*-H$_2$NPheNHPh	32–45	23	1.2 (D)	5.7	0.9

Source: Reproduced with permission from B. Sellegren, *Chirality*, **1**, 63 (1989). Copyright 1989 Wiley-Liss.

Disadvantages

They require large amounts of enantiomerically pure imprint molecules.

The efficiency is poor, though improvements are being made.

They are not commercially available.

Methodological considerations limit their usefulness.

Brush-type

These are prepared by immobilizing a small-molecular weight enantio-enriched selector on a chromatographic support (usually silica). Historically, mostly "chiral pool" selectors have been used. Totally synthetic selectors are being used now, and continuing improvements in selector design and immobilization chemistry are being made.

In these CSPs, π-π bonding interactions are important for the retention process. The interactions of this type are well known and have been reported to occur between π-donor and π-acceptor molecules. The earliest applications of the charge transfer type of complexation to optical resolution by HPLC was reported by Klemm et al. in 1960.[10, 11] A reagent based on π acidity has been designed and synthesized by Newman's group, where optically active α-(2,4,5,7-tetranitro-fluoroenylideneamino-oxy)propionic acid is found to be capable of resolving a variety of aromatic π bases by a charge transfer type of reaction.[12] Pirkle et al. made a significant breakthrough in charge-transfer based adsorption for optical resolution in HPLC by introducing N-(3,5-dinitrobenzoyl)amino acids as CSPs (Figure 6.10) for π-π interactions based on their work with anthryl carbinol.[13, 14] These types of CSPs can resolve a variety of racemic solutes, such as sulfoxides, amines, amino acids, hydroxy acids, lactones, alcohols, amino alcohols, and thiols (Table 6.5). The π-donor type of CSPs include structures such as hydantoin, arylalkylamine, and phthalide.

A list of commercially available brush-type phases is given in Table 6.6. These CSPs offer the following advantages and disadvantages:

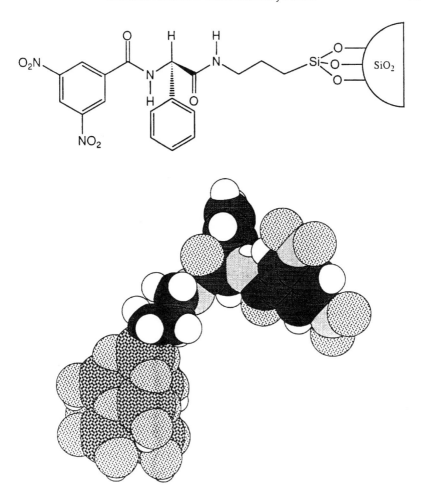

Figure 6.10 DNB-Phenylglycine CSP. Reproduced from S. Ahuja, W. Pirkle, and C. Welch, *Chiral Separations by Chromatography*, American Chemical Society Short Course, 1994. Copyright 1994 American Chemical Society.

Advantages

    High capacity
    High efficiency
    Stable
    Can invert elution order
    Good solvent compatibility
    Good for preparative separations

Table 6.5 Chromatographic Resolution of 3,5 Dinitrophenylated Derivatives of Some Amino Acids, Amines and Alcohols

Compound	Derivative[a]	k'_1 (enantiomer)	α	% 2-Propanol in Hexane
Normal Phase				
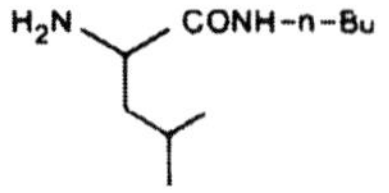 H₂N–CH(CH₂CH(CH₃)₂)–CONH-n-Bu	DNB	0.38 (R)	17.66	10
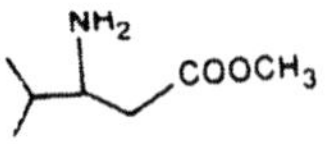 (CH₃)₂CH–CH(NH₂)–COOCH₃	DNB	0.93 (S)	1.97	5
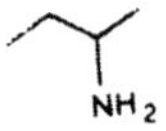 (CH₃)₂CH–CH(NH₂)–CH₃	DNAn	5.87 (S)	1.19	5
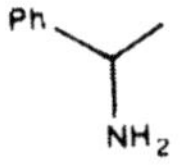 Ph–CH(NH₂)–CH₃	DNAn	3.27 (R)	1.33	20
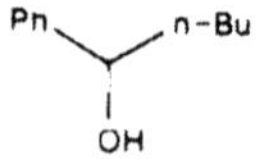 Ph–CH(OH)–n-Bu	DNAn	3.19 (R)	1.24	5
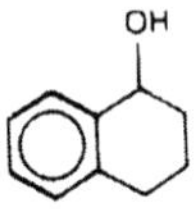 1,2,3,4-tetrahydronaphth-1-ol	DNAn	5.35 (R)	1.22	5
Reversed Phase				
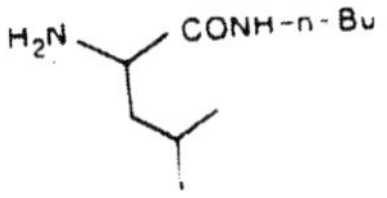 H₂N–CH(CH₂CH(CH₃)₂)–CONH-n-Bu	DNB	9.0 (R)	2.61	50% methanol-water

a. DNB = 3,5-dinitrobenzoyl-; DNAn = 3,5-dinitroanilido-.

CSP: -Si-(CH₂)₁₁-O-C(=O)-CH(i-Pr)-NH-C(=O)-(9-anthryl)

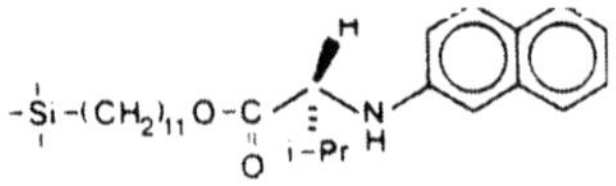

Source: Reproduced from W. H. Pirkle and T. C. Pochapsky, *J. Am. Chem. Soc.*, 108, 352 (1986). Copyright 1986 American Chemical Society.

Disadvantage

It is not possible to separate all enantiomers with the current selection of brush-type CSPs.

Comparison of Various CSPs

A comparative account of various CSPs is given in Table 6.7. A careful review of the table should be useful to readers in making a logical selection of a suitable CSP.

Table 6.6 Commercially Available Brush-Type Chiral Stationary Phases

Chiral Stationary Phase	Description/Absolute Configuration
(D)-Phenylglycine	(R)-N-(3,5-Dinitrobenzoyl)phenylglycine
(L)-Phenylglycine	(S)-N-(3,5-Dinitrobenzoyl)phenylglycine
(D)-Phenylglycine (ionic)	(R)-N-(3,5-Dinitrobenzoyl)phenylglycine
(D, L)-Phenylglycine	(R, S)-N-(3,5-Dinitrobenzoyl)phenylglycine
(D)-Leucine	(R)-N-(3,5-Dinitrobenzoyl)leucine
(L)-Leucine	(S)-N-(3,5-Dinitrobenzoyl)leucine
(D)-N2N-Naphthylalanine	(S)-N-(2-Naphthyl)alanine
(L)-N2N-Naphthylalanine	(R)-N-(2-Naphthyl)alanine
(D, L)-N2N-Naphthylalanine	(R,S)-N-(2-Naphthyl)alanine
(S)-N1N-Naphthyl leucine	(S)-N-(1-Naphthyl)leucine
α-Burke 1	(R)-Dimethyl N-3,5 dinitrobenzoyl-α-amino-2,2-dimethyl-4-pentyl phosphonate
β-Gem 1	(S, S)-N-3,5 dinitrobenzoyl-3-amino-3-phenyl-2-(1,1-dimethylethyl) propanoate
(*tert*)-Buc-(S)-leucine	N-(*tert*-Butylaminocarbonyl)-(S)-leucine
OA-1000	(S)-(1-Naphthyl)ethylamino teriphthalic acid
OA-2200	(R,3R)-*trans*-Chrysanthemoyl-(R)-phenylglycine
OA-2500	(R)-N-(3,5 dinitrobenzoyl)-1-naphthylglycine
OA-3000	N-(*tert*-butylaminocarbonyl)-(S)-valine
OA-3100	N-3,5 Dinitrophenylaminocarbonyl-(S)- valine
OA-3200	N-3,5 Dinitrophenylaminocarbonyl-(S)-(*tert*)-leucine
OA-4000	(R)-N-(1-Naphthyl)ethylaminocarbonyl-(S)-valine
OA-4100	(S)-N-(1-Naphthyl)ethylaminocarbonyl-(S)-valine
OA-4400	(S)-(1-Naphthyl)ethylaminocarbonyl-(S)-proline
OA-4500	(R)-(1-Naphthyl)ethylaminocarbonyl-(S)-proline
OA-4600	(R)-(1-Naphthyl)ethylaminocarbonyl-(S)-(*tert*) leucine
OA-4700	(S)-(1-Naphthyl)ethylaminocarbonyl-(S)-(*tert*)-leucine
(R)-N-1-Naphthylethylurea	(R)-1-(Naphthyl)ethylurea
(S)-N-1-Naphthylethylurea	(S)-1-(Naphthyl)ethylurea

Table 6.7 Comparison of Various CSPs

	Brush-type	Poly-saccharide	Protein	Cyclo-dextrin	LE	IP	Crown ether
Resolves enantiomers	+ +	+ +	+ +	+	+	+	+
High efficiency	+ +	+	+	−	+	−	+
Column stability	+ +	−	−	+	+	+	+
Mobile phase compatibility	+ +	−	+	+	−	+	+
Inversion of elution order	yes	no	no	no	no	yes	yes
High capacity	+ +	+	−	+	+	−	−
Analysis time	+	−	−	−	+	−	−

a. LE= Ligand exchange; IP = Imprinted polymer.

Source: Reproduced with permission from S. Ahuja, W. Pirkle, and C. Welch, *Chiral Separations by Chromatography*, American Chemical Society Short Course, 1994. Copyright 1994 American Chemical Society.

REFERENCES

1. S. Ahuja, W. Pirkle, and C. Welch, *Chiral Separations by Chromatography*, American Chemical Society Short Course, 1994.
2. G. M. Henderson and H. G. Rule, *Nature*, **141**, 917 (1938).
3. C. E. Dalgliesh, *Biochem. J.*, **52**, 3 (1952).
4. G. Hess and R. Hagel, *Chromatographia*, **6**, 277 (1973).
5. T. Shibata, I. Okamoto, and K. Ishi, *J. Liq. Chromatogr.*, **9**, 313 (1986).
6. S. Allenmark, *Chromatographic Enantioseparation*. Ellis Horwood: New York, 1991.
7. Y. Okamoto, K. Suzuki, K. Ohta, K. Hatada, and H. Yuki, *J. Am. Chem. Soc.*, **101**, 4763 (1979).
8. Y. Okamoto, K. Suzuki, and H. Yuki, *J. Polym. Sci., Polym. Chem. Ed.*, **18**, 3043 (1980).
9. B. Sellegren, *Chirality*, **1**, 63 (1989).
10. L. H. Klemm and D. Reed, *J. Chromatog.*, **3**, 364 (1960).
11. L. H. Klemm, K. B. Desai, and J. R. Spooner, *J. Chromatog.*, **14**, 300 (1964).
12. M. S. Newman, W. B. Lutz, and D. Lediner, *J. Am. Chem. Soc.*, **77**, 3420 (1955).
13. W. H. Pirkle and D. L. Sikenga, *J. Chromatog.*, **123**, 400 (1976).
14. W. H. Pirkle and T. C. Pochapsky, *J. Am. Chem. Soc.*, **108**, 352 (1986).

REVIEW

1. List various types of chiral stationary phases.
2. What significant advantage do brush-type stationary phases offer over other CSPs?
3. List one primary advantage and disadvantage of protein CSPs.
4. List one primary advantage and disadvantage of β-cyclodextrin-based CSPs.

TUTORIAL

The following list of desirable features for a chiral separation is quite useful in selection of a suitable CSP:

- Baseline resolution of enantiomers
- High efficiency
- Short analysis time
- Column stability (time, temperature, solvent)
- Mobile-phase compatibility
- Ability to invert elution order
- High capacity

For a comparative account of these features for a given CSP, review Table 6.7.

7

Understanding Chiral Chromatography

It is important to try to understand why separations occur on different chiral stationary phases. There is no denying the fact that at this stage much is still unknown as to the exact mechanism of separation. However, a review given below of our current knowledge of chiral separations should help the readers to carry out better separations and enable them to develop superior chiral phases. A number of theories abound. These are discussed here to generate some reasonable rationale for separations.

In general, separation scientists prefer to carry out resolution of enantiomers with minimal manipulations. Derivatization in this context is not considered favorably. Therefore separations that do not require derivatization are considered very useful for both analytical and preparative purposes. The greatest progress in this regard has been made in HPLC where derivatization is frequently unnecessary.

Enantioselective Interactions

Chromatographic separations generally occur because of adsorption, partition, ion exchange, or other molecular interactions relating to stereochemistry of the molecules. The observed separations relate to the type of stationary phase that is being used for a given separation. The chromatographic process is generally regarded as a series of equilibria between the stationary phase and the mobile phase. Introduction of bonded phases in GC and HPLC have made the distinction between adsorption- and partition-related separations quite difficult. However, the role of molecular interactions in the separation process cannot be underestimated.

It is reasonable to assume that separation of enantiomers on a chiral stationary phase (CSP) results from selective interactions with the stationary phase. The understanding as to the nature of these interactions remains a matter of speculation that deserves greater in-depth investigation. These investigations are continuing, and our current state of knowledge is summarized in this chapter.

A number of chiral recognition models have been proposed for resolution of optical isomers that are based on the three-point interaction theory advanced by Dalgliesh in 1952.[1] According to this theory, three simultaneous interactions are necessary between the enantiomer and the stationary phase. Dalgliesh reached this conclusion from the separation studies of some aromatic amino acids by paper chromatography. It was assumed that the hydroxyl groups of cellulose (paper) formed hydrogen bonds to the amino and carboxyl groups of the amino acid. The third interaction related to the aromatic ring substituents. This theory seems logical for the given separation and can be diagramatically represented as shown in Figure 7.1; however, there are many cases of enantiomeric separations where these type of interactions are not necessary. As a result, the separation occurs by other interactions (see Chromatographic Enantioselectivity).

The three-point interactions are frequently referred to as the three-points rule, a term popularized recently by Pirkle.[2] This approach considers chiral recognition as it might occur between small chiral molecules in solution. It is assumed that chiral recognition cannot occur unless there are at least three simultaneous interactions, at least one of which is stereochemically dependent, between a chiral "recognizer" and one of the enantiomers whose configuration is to be "recognized" (Figure 7.2).

The interactions thus employed are the usual interactions that occur in solution. It is necessary to design a chiral molecule that contains functional groups capable of undergoing these interactions, anchor it to a support, pack a column, and chromatograph a racemate that contains an appropriate functionality. Complementary functionality should be present if the analyte is to undergo the required multiple simultaneous interactions. This rationale has great value and far-reaching implications. It gives us an a priori basis for designing a chiral stationary phase (CSP), and provides a means of understanding the subsequently observed behavior of assorted racemates on the CSP. Furthermore, it offers a means of rationally improving the performance of a given CSP. Finally, and most importantly, when

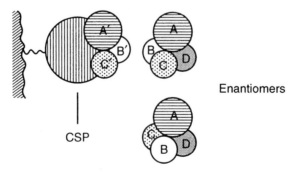

Figure 7.1 A cartoon of three-point interaction.

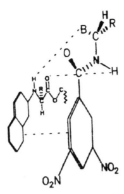

Figure 7.2   A chiral recognition model.

the CSP is prepared synthetically, its structure can be altered and controlled as desired.

The first CSP prepared in this series is 9-anthryl trifluoromethyl carbinol, linked at the 10-position to silica through a six-atom connecting arm. This CSP is intended to utilize π-π interactions, hydrogen bonding interactions, and steric interactions to separate the enantiomers of compounds containing π-acidic and basic sites.[3] Among the analytes that can be resolved are N-(3,5-dinitrobenzoyl) derivatives of amines and amino acids. Chiral recognition is reciprocal in that if a CSP derived from (+)-A selectively retains (+)-B, then a CSP derived from (+)-B should be selectively retained (+)-A (see Figure 7.3).

The structural requirements of these CSPs are predictable on the basis of a crit-

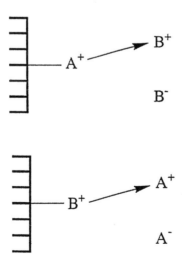

Figure 7.3   A diagrammatic representation of reciprocity.

ical review of resolved analytes, albeit somewhat imprecisely. For example, the N-(3,5-dinitrobenzoyl) amino acid columns use a combination of π-π hydrogen bonding, dipole-dipole, and steric interactions to achieve chiral recognition. Therefore, the analytes must contain a combination of π-basic, basic, and acidic sites; possibly have a strong dipole; or perhaps contain bulky steric interaction sites. These interaction sites need to act in concert.

Among the analyte enantiomers separable on the N-3,5-dinitrobenzoyl-phenylglycine CSPs are N-acylated arylalkylamines. To study the relationship between analyte structure, chromatographic separability, and operative chiral recognition mechanism(s), several reciprocal α-arylalkylamine-derived CSPs have been designed and prepared. These CSPs afford excellent selectivity for the N-(3,5-dinitrobenzoyl) derivatives of amines, α-amino esters, amino alcohols, and α-aminophosphonates.

Similar CSPs derived from linking α-arylalkylamines to silica through urea functionality have been prepared.[4,5] These CSPs are workable but are inferior to the corresponding amide-linked CSPs.[2] One important aspect of these CSPs is that they have more than one chiral recognition process available to them. It has been shown that these CSPs use two competing processes of opposite enantioselectivities.[6,7] Optimization requires altering the structure of the chiral entity to maximize the strength of essential interactions, the manner in which the chiral entity is connected to the silica support, and spacing between adjacent strands of the bonded phase.[8] Simply stated, one process is more intercalative than the other. Densely packed strands disfavor the intercalative process, thereby enhancing the contribution of the nonintercalative process. Similarly, the orientation of the chiral entity with respect to the silica surface determines whether, using a given combination of analyte-CSP interactions, a portion of the analyte is intercalated between adjacent strands. In an organic mobile phase, intercalation can lead to steric repulsion. Thus, a different orientation of chiral entity with respect to silica will alter the relative contributions of the two competing processes. High selectivity may be obtained by largely suppressing one of the processes. In aqueous mobile phases, lipophilic interaction begins to compensate for steric repulsion that attends intercalation. The practical consequence of the aqueous mobile phase is reduction in selectivity; however, selectivity is more than adequate for analysis of enantiomeric purity.

Chromatographic Enantioselectivity

We know that resolution of optical isomers in chromatography occurs because of reversible diastereomeric association between the chiral environment of a column and the solute enantiomers. Chromatographic optical resolution can be obtained under a variety of conditions. This suggests that the molecular association necessary for enantiomeric separations is possible by various types of molecular interactions. The association can be expressed in terms of the equilibrium constant, which relates attractive as well as repulsive interactions involved in the separation process. The attractive interactions include hydrogen bonding, electrostatic and dipole-dipole interactions, charge-transfer interaction, and hydrophobic interaction (as mentioned, they are more pronounced in aqueous systems). The repulsive interactions are usually steric; dipole-dipole repulsion can also occur.

Steric fit may be yet another significant route to enantiomeric separations. The types of CSPs where steric fit has been conjectured to be quite important are cyclodextrins and crown ether phases (see Inclusion). Although no CSP has yet been designed based exclusively on steric fit, molecular imprinting has been tried.[9] Steric selectivity or selective steric interactions may play a role in separations on protein-based columns (see Separations on Protein Columns); however, because of the complexity of these molecules, the exact sites that lead to enantiomeric separations have not been identified.

Hydrogen bonding alone has been shown to be an adequate source of separation for some enantiomers in GC, as well as in HPLC. The fact that an enantiomeric solute bearing one hydrogen binding substituent can be resolved under these conditions suggests that only one attractive force can suffice for chiral discrimination under certain chromatographic conditions. Therefore, it is desirable to explore several theoretical aspects of binding interactions that may be important to enantioselective separations.

Charge-Transfer Interactions

As mentioned, these interactions are commonly encountered with π-electron systems; notable examples of these systems have been provided by Pirkle. Charge-transfer interactions require formation of relatively stable charge-transfer complexes between aromatic rings that act as donors or acceptors. These π-π interactions, together with other polar interactions such as hydrogen bonding and dipole-dipole interactions, form the basis of efficient chiral selectors in HPLC.

Good π-electron donors are aromatics with electron-releasing substituents such as amino or alkoxy groups. On the other hand, nitroaromatics are generally quite good π-electron acceptors because the negative charge can be effectively delocalized by the participation of the substituents in resonance stabilization. The principle of this type of selectivity has been shown in various mixtures of organic solvents (e.g., 2-propanol and hexane) which frequently provide high α-values. Group matching of the two molecules leads to simultaneous interactions at three points, at least two of which are bonding (Figure 7.4).

Figure 7.4 depicts the retention of a strongly retained enantiomer. Evidence for these multiple interactions has been provided with data from intermolecular nuclear Overhauser effect in NMR and from molecular mechanics calculations.[11, 12] Additional support of the postulated mechanism comes from X-ray crystal structure analyses of analogous chiral charge-transfer complexes as represented in Figure 7.5.[13] The most important feature of these systems is reciprocity—the observed enantioselectivity is independent of the enantiomer (the molecule responsible for the formation of the complex) immobilized as the stationary phase on the column.

116   Chiral Separations by Chromatography

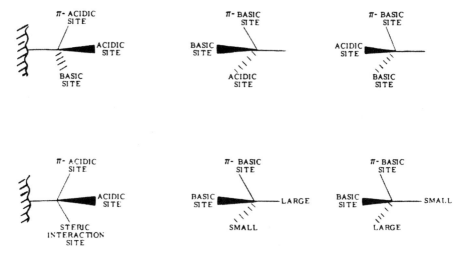

Figure 7.4 Stereoselectivity with charge-transfer chiral selectors.

Hydrogen Bonding

Diastereomeric complexes differing in free energy can be formed by intramolecular association via bidentate hydrogen bonding. Since bifunctional amides of this type are easily available from common amino acids as starting materials, it is understandable that the first successful separations in gas chromatography are based on this approach.[14] Carboxylic acid dimers provide two simultaneously equivalent hydrogen bonds that can exist even in the gaseous phase. A low polarity of the medium can improve the stability of this association, which makes this approach

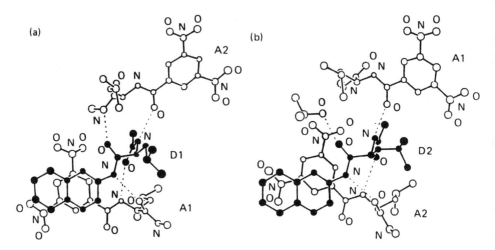

Figure 7.5 X-ray crystallographic representation of diastereomeric complexes.

useful for HPLC when nonpolar solvents are used.[15] A chiral selector can be used in the mobile phase or as a stationary phase by immobilizing it on the support. The diastereomeric complexes most likely behave as cyclic *cis-trans* isomers, and the steric differences cause different interactions with an achiral stationary phase in a chromatographic system. The dual hydrogen bond association and its importance to chiral recognition in GC and HPLC have been demonstrated by X-ray crystal structure determination on diastereomeric complexes.[16]

Temperature Effects

Recall the effect of temperature on chiroptical properties (see Specific Rotation in Chapter 5). Because the specific gravity of liquids, as well as optical activity of many substances, is influenced by temperature; it is necessary to indicate the temperatures at which rotation and specific gravity are determined.

The temperature can also have the following additional effects that can influence separations:

- Changes in the population of vibrational and rotational energy levels of the chiral solute
- Displacement of solute-solvent equilibria
- Displacement of conformational equilibria
- Aggregation and microcrystallization of the chiral solute

In general $[\alpha]$ changes 1–2% per degree C; however, larger changes have been observed, including a change in sign—for example, 0.5% solution of aspartic acid has $[\alpha]$ value of +4.4 at 20°C, 0 at 75°C, and −1.86 at 90°C.

GC separation of enantiomers is controlled by thermodynamics. On an optically active CSP, diastereomeric association complexes can be depicted as follows:[17]

$$CSP + A_R \overset{K_R}{\rightleftharpoons} CSP - A_R \qquad (7.1)$$

$$CSP + A_S \overset{K_S}{\rightleftharpoons} CSP - A_S \qquad (7.2)$$

A_R and A_S are analytes in R and S configurations.

If enantiomeric resolution is observed, the association constants K_R and K_S have different values. The free energy difference, $\Delta(\Delta G)$, can be calculated as follows:

$$-\Delta(\Delta G) = RT \ln \frac{K_R}{K_S} \qquad (7.3)$$

where R = gas constant and T = temperature in K.

Unfortunately, measuring K_R and K_S is not feasible in most cases. To a first approximation, $\Delta(\Delta G)$ can be calculated from the separation factor, α:

$$-\Delta(\Delta G) = RT \ln \alpha \qquad (7.4)$$

It must be recognized that retention of an enantiomer is determined both by gas-liquid equilibrium (achiral contribution) and diastereomeric association complexation (chiral contribution). $\Delta(\Delta G)$ is usually underestimated because the achiral contribution to retention cannot be totally eliminated. Table 4.7 shows some representative α values and $-\Delta(\Delta G)$ values obtained by GC and HPLC. It may be noticed that HPLC can provide higher values (see the following discussion).

Based on basic thermodynamic relationships, corresponding $\Delta(\Delta H)$ and $\Delta(\Delta S)$ can be obtained by measuring α values at different temperatures and then plotting R-ln-α versus $1/T$.

$$R \ln \alpha \; \frac{-\Delta(\Delta H)}{T + \Delta(\Delta S)} \qquad (7.5)$$

A straight line is obtained if $\Delta(\Delta H)$ is constant within a certain range; where the slope is $\Delta(\Delta H)$ and the intercept is $\Delta(\Delta S)$. The measurement of thermodyamic parameters for a number of compounds on derivatized cyclodextrin phases showed that, for most compounds, $\Delta(\Delta H)$ and $\Delta(\Delta S)$ have the same negative sign. This means there is an enantioselective temperature (T_{iso}) for these compounds. According to Equation 7.5, we get the following relationship at values of $\alpha = 1$:

$$T_{iso} = \frac{\Delta(\Delta H)}{\Delta(\Delta S)} \qquad (7.6)$$

At temperatures higher than T_{iso}, the enantiomer elution order should be reversed. However, for most compounds, T_{iso} is much higher than the working temperature range, and thus retention time of these compounds is equal to t_0 at T_{iso}. Figure 7.6 shows plots of enantioselectivity (α) versus T for four pairs of enantiomers on CSP, Octakis (2,6-di-O-n-pentyl-3-O-trifluoroacetyl) γ-cyclodextrin, that have similar T_{iso} and α value temperature relationships.

These plots show that enantioselectivity is highly temperature-dependent for compounds that have high absolute $\Delta(\Delta H)$ values. Temperature dependence is minimal for compounds with low $\Delta(\Delta H)$ values. This suggests that in compounds that have moderate to high $\Delta(\Delta H)$ values, it is advantageous to perform enantiomeric separations at lower column temperatures with short columns. Compounds with low $\Delta(\Delta H)$ values would require a long and highly efficient column or a more selective stationary phase.

The free energy difference associated with a given α value can be easily calculated from the chromatographic data:[10]

α	Δ(ΔG) (kcal/mol)
1.05	29
1.10	56
1.50	240
2.00	410

These data show that a very small energy difference is needed to resolve enantiomers on a reasonably efficient column. Capillary GC columns can give baseline resolution for peaks with α values of <1.05. For example, in a column with an effective plate number of 200,000, only 2% peak overlap is seen with an α value of 1.01 or resolution value of 1.11. This corresponds to an energy difference of only 5.9 kcal/mol. These low values are at least an order of magnitude lower than those values associated with conformational changes in a molecule. Apparently the interactions of the two enantiomers to a given chiral site may involve different amounts of energy, because for steric reasons, one of the enantiomers might be forced to adopt energetically less favorable conformation.

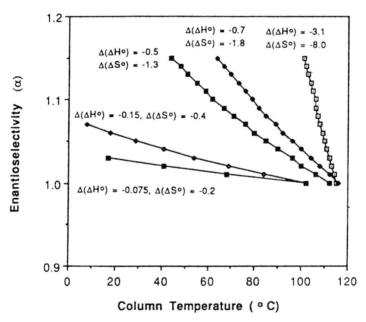

Figure 7.6 Plots showing the effect of temperature on enantioselectivity of compounds on a Chiraldex G-TA column. Reproduced with permission from W. Li and T. M. Rossi, in *The Impact of Stereochemistry on Drug Development and Use*, H. Y. Aboul-Enein and I. W. Wainer, Eds., Wiley, New York, 1997. Copyright 1997 John Wiley and Sons.

Column efficiency is generally much lower in HPLC; however, this is easily offset by high α values offered by the technique. It is possible to find α values of 30 or more that correspond to $\Delta(\Delta G)$ values of 2 kcal/mol. Such high α values are generally obtained due to low retention of one of the enantiomers. This suggests that a very selective enantioselective process is operating in the column; in other words, one of the enantiomers is showing minimal interactions for steric reasons.

Column temperature has been given very little attention in most HPLC separations, since the separations are generally conducted at room temperature.[18] Variable temperature runs can provide useful information concerning the thermodynamic parameters for the CSP-analyte interactions. An example of such a separation is shown in Figure 7.7. The results clearly show that at the lower end of the investigated temperature range, the capacity factor (k') of both isomers and λ values increase with a small increase in peak width. The peak broadening may be due to a slow mass transfer process occurring in the investigated temperature range. However, it is clear that decreasing the temperature causes an enhancement of the chromatographic resolution of the peaks corresponding to both enantiomers. These results suggest that it is important to evaluate various CSPs individually to study the effect of temperature with the analytes of interest.

Inclusion

Starch is known to include iodine within its structure. Crystalline degradation products of starch—cyclodextrins obtained by the action of microorganisms—have found use in chromatographic separation of enantiomers. For example, α-cyclodextrin is composed of a ring with six glucose units, which has the right size for inclusion of benzene. Interestingly, β-cyclodextrin (composed of seven glucose

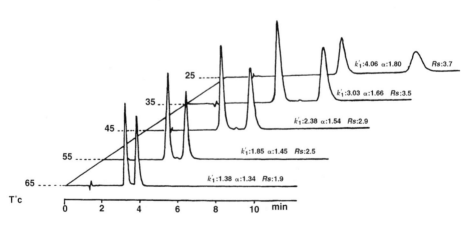

Figure 7.7 Temperature influence on k' α and resolution of an amido compound. Reproduced with permission from F. Gasparrini, D. Misiti, and G. Villani, in *Recent Advances in Chiral Separations*, D. Stevenson and I. D. Wilson, Eds., Plenum, New York, 1990. Copyright 1990 Plenum Publishers.

units) is precipitated by bromobenzene as a consequence of inclusion complex formation. The steric requirements of this host-guest complex formation imply that this phenomenon is highly stereoselective. This observation suggested that by the use of a chiral host it would be possible to separate chiral molecules. A large number of HPLC separations of enantiomers have been reported that have used this concept in approaching resolutions (see chapter 10).

Researchers believe that the cavity is hydrophobic and is, therefore, suitable for inclusion of the hydrocarbon portion of the molecule. No bonds are involved, as the inclusion is merely a result of the hydrophobic effects, and the structural demands are not pronounced except as dictated by the size of the cavity. In reversed-phase HPLC, the combination of hydrophobic interactions, which encourage inclusion, and steric effects from the substituents present in a chiral structure on the cavity entrance are believed to be responsible for the observed enantioselectivity.

Macrocyclic antibiotics that were developed as an extension of the inclusion process[19] have been found to be useful. These stationary phases appear to be multimodal in that they can be used both as normal and reversed-phase modes. Enantioseparation is possible via several different mechanisms, including π-π complexation, hydrogen bonding, inclusion in a hydrophobic pocket, dipole stacking, steric interactions, and combinations thereof.

Chiral crown ethers show remarkable enantioselectivity toward organic ammonium ions. In this case, the ammonium ion is held to other oxygen atoms within the hydrophobic cavity by hydrogen bonds (Figure 7.8). Clearly, the structural and steric requirements of the guest molecule are much higher than required by cyclodextrins.

The inclusion effects seen in chiral matrices composed of swollen microcrys-

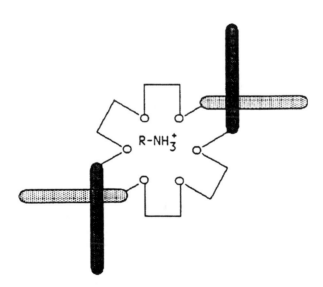

Figure 7.8 A cartoon of a crown ether CSP.

talline cellulose are interesting for a number of reasons. Cellulose triacetate, prepared by heterogeneous acetylation to preserve microcrystallinity, has been shown to act partially by steric exclusion effects. Investigations show that benzene is highly retained on this phase, whereas mesitylene is much less retained and 1,3,5-tri-tertiary butyl benzene is not retained. This phenomenon may be explained by considering the lamellar arrangement of polysaccharide chains, which might produce a two-dimensional molecular sieve that allows inclusion of flat aromatic molecules but excludes sterically more demanding structures. The higher retention of benzene as compared to toluene further supports the suggestion of pockets in the chain structure.

Transition Metal Complexes

A transition metal complex can be formed by a ligand, which may donate electrons to the unfilled inner shell *d*-orbitals. The coordination complexes thus formed possess a well-defined geometry such that ligands can occupy certain given positions in space. The donor ligand atoms in the complex are held at strictly fixed distances from the metal atom in defined orientations (commonly referred to as a coordination sphere). The stability of the complex should be stereochemically dependent because if two or more chiral ligands are present in the coordination sphere, their mutual interaction either directly or through the solvent molecule of the first or second sphere will provide differences in complex stability and thus provide enantioselectivity.

To utilize this approach in chromatography, it is necessary that the complexes formed should be sufficiently kinetically labile. That is, their formation and dissociation should be sufficiently fast to meet the chromatographic time scale. The complex stability is highly dependent on the transition metal used; generally, Cu(II) complexes are the most stable and preferred in HPLC, whereas the less stable complexes formed with Ni(II), Co(II), and Zn(II) have been studied in GC.

Both forms of chromatography use a chiral stationary phase composed of an immobilized chiral ligand forming a coordination complex with the transition metal ion. Reversible diastereomeric mixed ligand sorption complexes formed by a displacement or exchange mechanism lead to the separation of a suitable racemic mixture injected onto a chromatographic column.

Separations on Protein Columns

Significant progress has been made in the resolution of chiral compounds on protein-based columns. This progress relates largely to the improvements in immobilization technology and production of cross-linked silica-bonded phases that are compatible with organic solvents. Protein columns are often called biopolymeric phases, which are more likely to be quite complex structures that exhibit broad enantioselectivity and prove to be useful for separation of enantiomers of a variety of racemates, particularly of polar or charged compounds, as represented by many pharmaceutical compounds.

In 1958, it was observed that L-tryptophan has 100 times higher affinity for bovine serum albumin (BSA) than the D-enantiomer does. Over two decades later, in 1982, it was demonstrated that low-pressure LC on a BSA column can separate aromatic amino acids, a chiral sulfoxide and sulfimide, and N-benzoylamino acid. This then led to a number of protein columns:

- α1-Acid glycoprotein column (AGP) from orosomucoid
- Ovomucoid column, OVM, an acid glycoprotein from chicken egg white
- Cellobiohydralase-I (CBH-I), from a microbial process (mold-*Trichoderma reesci*)

To improve our understanding of these columnns, their binding to silica is summarized below.

Glycoproteins (AGP, OVM, CBH-I) on oxidation with periodate yield aldehyde groups in the carbohydrate moiety, which then react with the primary amino group of the proteins, giving easily reducible imino functions. If the silica is activated to contain either amino or aldehyde groups, the process will simultaneously anchor the cross-linked protein to silica. Albumin requires a cross-link, a bifunctional agent such as glutaraldehyde and N,N'-disuccinimmidyl carbonate.

The following are some of the important requirements of immobilization techniques:

- It is necessary to have a high degree of binding to silica.
- Immobilization techniques invariably contribute to non-stereoselective binding.
- Since hydrophobic interaction makes a major contribution to retention on protein columns, any additional contributions from achiral structural elements of the stationary phase should be avoided.
- Free amino groups in the protein (probably those of lysine side chains) are coupled to suitably activated silica.
- Stability of the sorbent is increased if the protein is cross-linked during the immobilization procedure.

Protein columns can be used for separations of cationic or anionic solutes: BSA-type columns are useful for anionic analytes (carboxylic acids); AGP-type columns are useful for cationic analytes (amines); and OVM and CBH-I columns are useful for cationic analytes. It is important to remember that the sugar part of the molecule may be essential for chiral recognition.

The characteristics of some of the proteins used as stationary phases for chiral separations are given in Table 7.1. Protein binding chiral discrimination by HPLC stationary phases made with whole, fragmented, and third-domain turkey ovomucoid have been investigated.[20] Individual protein domains and two domains in combination have been prepared by enzymatic and chemical cleavage of turkey ovomucoid, followed by isolation and purification by size exclusion and ion exchange chromatography. The amino acid sequence of turkey ovomucoid is shown in Figure 7.9.[21]

Table 7.1 Properties of Protein Columns

Property	Orosomucoid	Ovomucoid	BSA
Molecular weight	41,000	28,800	66,000
Isoelectric point	2.7	3.9–4.5	4.7
Disulfide bridges	2	8	17
% Carbohydrate	45	30	

Silica-bonded phase HPLC columns were made from either whole or isolated domains of turkey ovomucoid. The columns made from whole turkey ovomucoid displayed chiral activity toward many racemates, whereas a combination of the first and second domains resolved only a selected number of aromatic weak bases. The first and second domains independently gave no appreciable chiral activity. The turkey ovomucoid third domain exhibited enantioselective protein binding fused-ring aromatic weak acids. Titration of third domain with model compounds, in con-

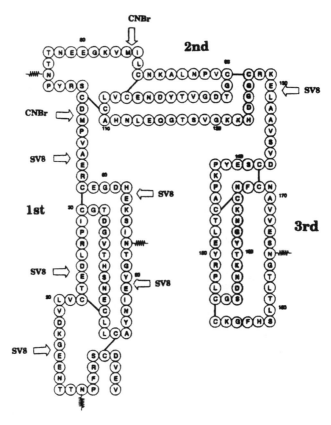

Figure 7.9 Amino acid structural sequence of turkey ovomucoid.

junction with NMR measurements, enabled the identification of amino acids responsible for binding. Molecular binding of the ligand-protein complexation indicates the ability of the protein surface to discriminate enantiomers on the basis of multiple molecular interactions.

Molecular Modeling

Advances in computer technology have increased our interest in the application of computational methods to the problem of chiral discrimination. These methods are based on evaluation of the potential energy contours obtained in molecular docking processes. The energy difference between the two complexes (at their potential energy minima) can be calculated by computerized matching of the contours found from docking of two enantiomeric molecules with the same chiral entity. This offers an approach to predict the elution order mechanism in a chromatographic system. The solvent effects cannot be adequately considered with this approach, and this limits the predictive value. Furthermore, even in a very well-defined chromatographic system containing a simple chiral selector, different retention mechanisms may compete,[22] thus making the situation more complex than is generally assumed in the theoretical approach. However, data from computational work on chiral discrimination are interesting and should be pursued to improve our understanding of molecular interactions affecting the mechanism of differential sorption observed in chromatography. In this context, computer-aided models for optimization of eluent parameters[23] and column selection[24] may be useful.

REFERENCES

1. C. Dalgliesh *J. Chem. Soc.*, 137, 1952.
2. W. H. Pirkle, In *Chromatography and Separation Chemistry*, S. Ahuja, Ed. American Chemical Society: Washington, DC, 1986, p. 101.
3. W. H. Pirkle, D. W. House, and J. M. Finn, *J. Chromatogr.*, **192**, 143 (1980).
4. N. Oi, M. Nagase, and T. Doi, *J. Chromatogr.*, **257**, 111 (1983).
5. W. H. Pirkle, M. H. Hyun, *J. Chromatogr.*, **322**, 295 (1985).
6. W. H. Pirkle, M. H. Hyun, and B. Bank, *J. Chromatogr.*, **316**, 585 (1984).
7. W. H. Pirkle, M. H. Hyun, A. Tsipouras, B. C. Hamper, and B. Bank, *J. Pharm. Biomed. Anal.*, **2**, 173 (1984).
8. W. H. Pirkle, C. J. Welch, and M. H. Hyun, *J. Org. Chem.*, **48**, 5022 (1983).
9. B. Sellegren, *Chirality*, **1**, 63 (1989).
10. S. Allenmark, *Chromatographic Enantioseparation*. Ellis Horwood: New York, 1991.
11. W. H. Pirkle and T. C. Pochapsky, *J. Am. Chem. Soc.*, **108**, 5627 (1986).
12. K. B. Lipkowitz, D. A. Demter, C. A. Parish, and T. Darden, *Anal. Chem.*, **59**, 1731 (1987).
13. R. Dappen, G. Rihs, and C. W. Mayer, *Chirality*, **2**, 185 (1990).
14. E. Gil-Av and B. Feibush, *Tetrahedron Lett.*, 3345 (1967).
15. A. Dobashi and S. Hara, *Tetrahedron Lett.*, 1509 (1983).
16. Y. Dobashi, S. Hara, and Y. Iitaka, *J. Org. Chem.*, **53**, 3894 (1988).
17. W. Li and T. M. Rossi, in *The Impact of Stereochemistry on Drug Development and Use*, H.Y. Aboul-Enein and I. W. Wainer, Eds. Wiley: New York, 1997, p. 415.

18. F. Gasparrini, D. Misiti, and G. Villani, in *Recent Advances in Chiral Separations*, D. Stevenson and I. D. Wilson, Eds. Plenum: New York, 1990, p. 109.
19. D. W. Armstrong, Y. Tang, S. Chen, Y. Zhou, C. Bagwill, and J-R. Chen, *Anal. Chem.*, **66**, 1473 (1994).
20. T. C. Pinkerton, W. J. Howe, E. L. Ulrich, J. P. Cominskey, J. Haginaka, T. Murashima, W. F. Walkenhorst, W. M. Westler, and J. L. Markley, *Anal. Chem.*, **67**, 2354 (1995).
21. I. Kato, J. Schrode, W. Kohr, and M. Laskowski, *Biochem.*, **26**, 193 (1987).
22. W. H. Pirkle and R. Dappen, *J. Chromatogr.*, **404**, 107 (1987).
23. R. P. Tucker, A. F. Fell, J. C. Berridge, and M. W. Coleman, *Chirality*, **4**, 316 (1992).
24. S. T. Stauffer and R. E. Dessy, *J. Chromatogr. Sci.*, **32**, 228 (1994).

REVIEW

1. What is a Dalgliesh interaction model?
2. Describe charge-transfer interactions.
3. How does inclusion affect chiral separations?
4. Describe the separation mechanism of protein columns.

TUTORIAL

Although much is still unknown as to the exact mechanism of separation on chiral stationary phases, it is important to understand why separations occur on different phases. We can reasonably assume that separation of enantiomers on a chiral stationary phase (CSP) results from selective interactions with the stationary phase. The understanding as to the nature of these interactions remains a matter of speculation that deserves in-depth investigation.

A number of chiral recognition models have been proposed for resolution of optical isomers that are based on the three-point interaction theory advanced by Dalgliesh in 1952. According to this theory, three simultaneous interactions are necessary between the enantiomer and the stationary phase. The three-point interactions are frequently referred to as the three-points rule, and this has recently been popularized by Pirkle. This approach considers chiral recognition as it might occur between small chiral molecules in solution. It is assumed that chiral recognition cannot occur unless there are at least three simultaneous interactions, at least one of which is stereochemically dependent, between a chiral "recognizer" and one of the enantiomers whose configuration is to be "recognized."

We know that resolution of optical isomers in chromatography occurs because of reversible diastereomeric association between the chiral environment of a column and the solute enantiomers. Chromatographic optical resolution can be obtained under a variety of conditions. This suggests that the molecular association necessary for enantiomeric separations is possible by various types of molecular interactions. The association can be expressed in terms of the equilibrium constant, which relates attractive as well as repulsive interactions involved in the separation process. The attractive interactions include hydrogen bonding, electrostatic and dipole-dipole interactions, charge-transfer interaction, and hydrophobic interaction (these are more pronounced in aqueous systems). The repulsive interactions are usually steric; dipole-dipole repulsion can also occur.

Hydrogen bonding alone has been shown to be an adequate source of separation for some enantiomers in GC as well as in HPLC. The fact that an enantiomeric solute, bearing one hydrogen binding substituent, can be resolved under these conditions suggests that only one attractive force is sufficient for chiral discrimination under certain chromatographic conditions.

Inclusion has been proposed as a possible mechanism for separations on cyclodextrin columns because the cavity is hydrophobic and is, therefore, suitable for inclusion of the hydrocarbon portion of the molecule. No bonds are involved, as the inclusion is merely a result of the hydrophobic effects, and the structural demands are not pronounced except as dictated by the size of the cavity. In reversed-phase HPLC, the combination of hydrophobic interactions, which encourage inclusion, and steric effects from the substituents present in a chiral structure on the cavity entrance is believed to be responsible for the observed enantioselectivity.

Transition metal complexes can be used for resolution of enantiomers. A transition metal complex can be formed by a ligand, which may donate electrons to the unfilled inner shell d-orbitals. It is necessary that the complexes formed should be sufficiently kinetically labile—that is, their formation and dissociation should be sufficient to meet the chromatographic time scale. The coordination complexes thus formed possess a well-defined geometry such that ligands can occupy certain given positions in space. The donor ligand atoms in the complex are held at strictly fixed distances from the metal atom in defined orientations (commonly referred to as a coordination sphere). The stability of the complex should be stereochemically dependent because if two or more chiral ligands are present in the coordination sphere, their mutual interaction, either directly or through the solvent molecule of the first or second sphere, will provide differences in complex stability and thus provide enantioselectivity.

Separations on protein columns are not well understood because of the complex nature of these molecules. It is generally believed that molecular binding of the ligand-protein complexation indicates the ability of the protein surface to discriminate enantiomers on the basis of multiple molecular interactions.

8

Method Development

A mixture consisting of equal amounts of enantiomers, called a racemate, is obtained experimentally when chemical reactions are carried out under a nonchiral environment. This necessitates the separation of the enantiomeric mixture—that is, an optical resolution is necessary to obtain optically pure species. As previously mentioned, enantiomers have identical physical properties except in terms of the rotation of the plane of polarized light. The enantiomer that rotates light to the right is referred to as dextro or d-enantiomer, and this type of rotation is written with a plus sign. By corollary, the other enantiomer is called levo or l-enantiomer and given a negative sign. It is important to separate these enantiomers to study their chiroptical and biological properties and to control their purity.

Since Louis Pasteur reported the first example of optical resolution in 1848, more than 1,000 compounds have been resolved, mainly by fractional recrystallization of diastereomeric salts. Resolution by entrainment or with enzymes or bacteria has also been applied to large-scale separation of some racemates, for example, amino acids. A review of various available methods for resolution of enantiomers reveals that more chiral compounds have been reported to be resolved by chromatographic methods than by crystallization.[1-7]

Stereoisomeric Interactions

It is now recognized that chromatographic methods offer distinct advantages over classic techniques in the separation and analysis of stereoisomers, particularly for the more difficult class of enantiomers.[1-19] Many factors are responsible for the extent of interactions of stereoisomeric molecules in any environment. A number of these factors are summarized in the following list. The nature and effects of some of these factors can influence chromatography of stereoisomers and must be carefully reviewed before developing a separation method.

Dipole-dipole interactions	pK differences — extent of ionization
Electrostatic forces	Resonance interactions/stabilization
Hydrogen bonding	Solubilities
Hydrophobic bonding	Steric interference (size, orientation, and spacing of groups)
Inductive effects	
Ion-dipole interactions	Structural rigidity/conformational flexibility
Ligand formation	Temperature
Partition coefficient differences	van der Waals forces
π-π interactions	

As discussed in chapter 7, an understanding of the mechanism of chiral separation is important in the development of optimum methods. Our understanding of chiral separations with some of the systems is quite good, but remains poor for protein and cellulose stationary phases.

The chromatographic process can be considered a series of equilibria, and the equilibrium constant describes the distribution of a compound between the stationary and mobile phases. Separation occurs because of differences in retention of various compounds on the stationary phase. It is customary to distinguish between partition and adsorption chromatography, in which the stationary phase is a liquid for the former and a solid for the latter. With the introduction of bonded organic phases in HPLC, this distinction is not very clear. Therefore, primary importance should be given to the types of molecular interactions with the stationary phase that lead to retention.

In adsorption chromatography, some kind of bonding interaction with the sorbent must occur. This may arise through a noncovalent attachment possible under the prevailing conditions. For example, hydrogen bonding, as well as ionic or dipole attraction, is enhanced by nonpolar solvents, whereas hydrophobic interactions are more important in aqueous media.

A number of chiral recognition models have been proposed to account for optical resolutions by HPLC, which are often based on the three-point interaction rule advanced by Dalgliesh[20] in 1952. He arrived at his conclusions from paper chromatographic studies of certain aromatic amino acids. It is assumed that the hydroxyl groups of the cellulose are hydrogen-bonded to the amino carboxyl groups of the amino acid. A third interaction is caused, according to these views, by the aromatic ring substituents. This led to the postulation that three simultaneously operating interactions between an enantiomer and the stationary phase are needed for chiral discrimination. However, this is not always necessary. Steric discrimination could also result from steric interactions.

Chiral separations are also possible through reversible diastereomeric association between a chiral environment, introduced into a column, and an enantiomeric solute. Since chromatographic resolutions are possible under a variety of conditions, one can conclude that the necessary difference in association can be obtained by many types of molecular interactions. The association, which may be expressed quantitatively as an equilibrium constant, will be a function of the magnitudes of the binding, as well as the repulsive interactions involved. The latter are usually steric, although dipole-dipole repulsions may also occur. Various kinds of binding interactions may operate, including hydrogen bonding, electrostatic and dipole-dipole attractions, charge-transfer interaction, and hydrophobic interaction (in aqueous systems). At times, a single type of bonding interaction may be sufficient to promote enantiomer differentiation. For example, hydrogen bonding, as the sole source of attraction, is sufficient for optical resolution in some HPLC modes of separation. The fact that enantiomeric solutes bearing only one hydrogen bonding substituent can be separated under such conditions supports the conclusion that only one attractive force is necessary for chiral discrimination in this type of chromatography. Taking a one-point binding interaction as a model, we may envision a difference in the equilibrium constant of the two enantiomers at the chiral binding site as due to effects from the site, forcing one of the enantiomers to acquire an unfavorable conformation. This resembles a situation often assumed to be present in enzyme-substrate interactions to account for substrate specificity.

CSPs, where steric fit is of primary importance, include those based on inclusion phenomena, such as cyclodextrin and crown ether phases. It may be possible to construct chiral cavities for the preferential inclusion of only one enantiomer. Molecular imprinting techniques are very interesting in this respect.[21] The idea is to create rigid chiral cavities in a polymer network in such a way that only one of two enantiomers will find the environment acceptable.

Chromatographic Methods

Chromatographic methods show promise for moderate-scale separations of synthetic intermediates, as well as for final products. For large-scale separations and in consideration of the cost of plant-scale resolution processes, the sorption methods offer substantial increases in efficiency over recrystallization techniques. The latter are still more commonplace because of the limited extent of current knowledge about stereospecific reactions required to tailor such separations.

Of various stereoisomers, diastereomers specifically are inherently easier to separate because they already possess differences in physical properties. In recent years, many significant advances have occurred that allow the chromatographic resolution of enantiomers. There are basically two approaches to the separation of an enantiomer pair by chromatography. In the indirect approach, the enantiomers may be converted into covalent, diastereomeric compounds by a reaction with a chiral reagent, and these diastereomers are typically separated on a routine, achiral stationary phase. In the direct approach, some variations can be tried: (a) the enantiomers or their derivatives may be passed through a column containing a chiral stationary phase or (b) the derivatives may be passed through an achiral column using a chiral solvent or, more commonly, a mobile phase that contains a chiral additive. Both variants of the second case depend on differential, transient diastereomer formations between the solutes and on the selector to effect the observed separation.

As previously mentioned, chromatographic methods such as TLC, GLC, and HPLC offer distinct advantages over classic techniques in the separation and analysis of stereoisomers, especially enantiomers that are generally much more difficult to separate.[4,15–19] Most of the discussion in this chapter is on HPLC, as it offers the greatest promise in this area.

Because HPLC is now one of the most powerful separation techniques, resolution of enantiomers by HPLC is moving rapidly with the availability of efficient chiral stationary phases. Large-scale, preparative liquid chromatographic systems have already been put on the market as process units for isolating and purifying chemicals and natural products. Chiral HPLC is ideally suited for large-scale preparation of optical isomers.

Derivatization of a given enantiomeric mixture with a chiral reagent, leading to a pair of diastereomers (indirect method), allows separation of samples by chromatography. On the other hand, using the chiral stationary or mobile-phase systems in chromatography (direct method) can provide a useful alternative procedure. This approach has been examined rather extensively by many research scientists. Early successful results did not attract much interest; the technique remained relatively dormant, and little was done to develop this approach into a generally applicable method. Approximately 30 years ago, systematic research was initiated for the design of chiral stationary phases functioning to separate enantiomers by gas chromatography. Molecular design and preparation of chiral phase systems for liquid chromatography have been well examined since then. More recently, efforts have been directed to finding new types of chiral stationary and mobile phases on the basis of the stereochemical viewpoint and the technical evolution of modern liquid chromatography.

Modes of Separation in HPLC

The chromatographic separation of enantiomers can be achieved by various methods; however, it is always necessary to use some kind of chiral discriminator or selector.[22,23] Two different types of selectors can be distinguished: a chiral additive in the mobile phase or a chiral stationary phase (discussed in this chapter). Another possibility is precolumn derivatization (discussed in this chapter) of the sample with chiral reagents, to produce diastereomeric molecules that can be separated by chiral chromatographic methods.

Some discussion on the mechanism of separation is provided for each mode of separation; however, as pointed out earlier, detailed mechanisms for chiral separations have not been worked out. The proposals made by certain scientists appear attractive; however, vigorous differences prevail, so no single proposal is emphasized here. The following discussion details the various approaches that can be used for chiral separations.

Chromatography of Diastereomeric Derivatives

Chromatography of diastereomeric derivatives is the oldest and most widely used approach for the resolution of enantiomers.[6] The precolumn derivatization of an optically active solute with another optically active molecule depends on the ability to derivatize the target molecule. A large number of functional groups and derivatives (shown in parentheses) have been investigated, including amino groups (amides, carbamates, ureas, thioureas, and sulfonamides), hydroxyl groups (esters, carbonates, and carbamates), carboxyl groups (esters and amides), epoxides (isothiocyanate), olefins (chiral platinum complexes), and thiols (thioesters).

This approach has been used with a wide variety of HPLC columns and mobile phases, including normal- and reversed-phase approaches. At present, there is no definitive way of determining which chromatographic approach will work.

Advantages

> The methodology has been extensively studied, making the application relatively easy and accessible.
>
> It is possible to use readily available standard HPLC supports and mobile phases.
>
> Detectability can be improved by appropriate selection of a derivatizing agent with a strong chromophore or fluorophore.

Disadvantages

> The synthesis of the diastereomeric derivatives requires initial isolation of the compounds of interest prior to their derivatization. This hinders the development of an automated procedure for large numbers of samples.
>
> The application to routine assays often is limited by enantiomeric contamination of the derivatizing agent, which can lead to inaccurate determinations. The problem of enantiomeric contamination of the derivatizing agent has been encountered in a number of studies. Silber and Riegelman,[24] for example, used (−)-N-trifluoroacetyl-L-prolyl chloride (TPC) in the determination of the enantiomeric composition of propranolol in biological samples. They found that commercial TPC was contaminated with 4% to 15% of the (+)-enantiomer and that the reagent rapidly racemized during storage.
>
> Enantiomers can have different rates of reaction and/or equilibrium constants when they react with another chiral molecule. As a result, two diastereomeric products may be generated in proportions different from the starting enantiomeric composition.[25]

Enantiomeric Resolution using Chiral Mobile-Phase Additives

Resolution of enantiomeric compounds has been accomplished through the formation of diastereomeric complexes with a chiral molecule(s) added to the mobile phase. Chiral resolution is due to differences in the stabilities of the diastereomeric complexes, solvation in the mobile phase, or binding of the complexes to the solid support. A general overview of this method has been published by Lindner and Pettersson.[26]

There are three major approaches to the formation of diastereomeric complexes: transition metal ion complexes (ligand exchange), ion pairs, and inclusion complexes. Each method is based on the formation of reversible complexes and uses an achiral chromatographic packing.

Ligand Exchange

Chiral ligand-exchange chromatography is based on the formation of diastereomeric complexes involving a transition metal ion (M), a single enantiomer of a chiral molecule (L), and the racemic solute (D and L). The diastereomeric mixed chelate complexes formed in this system are represented by the following formulas: L-M-D and L-M-L. The most common transition metal ion used in these separations is Cu, and the selector ligands are usually amino acids such as L-proline. The chromatography is most often carried out using an achiral HPLC packing (such as C-18), with these compounds added to the mobile phase.

The efficiency and selectivity of a chiral ligand-exchange system can be improved by binding the selector ligand to the stationary phase. Some examples of this approach are the L-proline-containing stationary phase, an L-(+)-tartaric-acid-modified silica reported by Kicinski and Kettrup, and a chiral phase composed of a C-18 column dynamically coated with (R,R)-tartaric acid mono-n-octylamide.[6]

A number of chiral molecules have been resolved by ligand-exchange chromatography. However, the resolution is possible for only those molecules that are able to form coordination complexes with transition metal ions. This method is most often utilized with free and derivatized amino acids and similar compounds. There has been some success with other classes of compounds, including carboxylic acids, amino alcohols (as Schiff bases), barbiturates, hydantoins, and succinimides.[26] The mobile phases employed with chiral ligand exchange are aqueous, with the metal ions and selector ligands added as modifiers.

Chiral ligand exchange is an excellent method for resolution of amino acids and amino acid-like compounds. The molecules need not be derivatized, and the aqueous mobile phases are compatible with automated column-switching techniques.

The major disadvantage of chiral ligand exchange is the small number of compounds that can be resolved by this approach. Many of the cationic and anionic molecules of pharmacologic interest have not been successfully resolved by this method.

Ion Pairing

Ion-pair chromatography is a liquid chromatographic method commonly used with charged solutes. The method is based on the formation of a neutral complex (ion pair, SC) between a charged solute (S) and a counterion of opposite charge (C).

When both the solute and the counterion are optically active, diastereoisomeric ion pairs are formed. These ion pairs often can be separated by differences in their solvation in the mobile phase or in their binding to the stationary phase. A number of different counterions have been employed in this approach, including (+)-10-camphor-sulfonic acid, quinine, quinidine, cinchonidine, (t)di-*n*-butyltartrate, and the protein albumin. This method has been reviewed by Pettersson and Schill.[27] The solutes resolved by chiral ion-pair chromatography have included amino alcohols such as alprenolol, carboxylic acids such as tropic acid and naproxen, and amino acids such as tryptophan.

The composition of the mobile phase depends on the chiral agent used. When a chiral counterion is used, a mobile phase of low polarity such as methylene chloride is used to promote a high degree of ion-pair formation. The retention of the solute can be decreased by increasing the concentration of the counterion or by the addition of a polar modifier such as 1-pentanol.[27]

The latter approach usually results in a decrease in stereoselectivity. The water content of the mobile phase also appears to be important, and a water content of 80 to 90 ppm has been recommended. With serum albumin as the chiral agent, aqueous mobile phases containing phosphate buffers are used. Retention and stereoselectivity can be altered by changing the pH. Both aqueous and nonaqueous mobile phases can be used when (t)di-*n*-butyltartrate is the chiral modifier. In some cases, it appears that the modifier is retained by the stationary phase when the column is equilibrated with an aqueous mobile phase. The system then can be used with an organic mobile phase[27].

Chiral ion-pair systems are not stable. The chromatography can be affected by the water content of the mobile phase, temperature, pH, and a number of other variables. This makes routine applications difficult. In addition, the counterions often absorb in the UV region, reducing the sensitivity of the system; indirect photometric detection[28] or other detection methods must be used.

Inclusion

Cyclodextrins are cyclic oligosaccharides composed of α-D-glucose units linked through the 1,4 position. The three most common forms of this molecule are α-, β-, and γ-cyclodextrin, which contain six, seven, and eight glucose units. Cyclodextrin has a stereospecific, doughnut-shaped structure. The interior cavity is relatively hydrophobic, and a variety of water-soluble and -insoluble compounds can fit into it, forming inclusion complexes. If these compounds are chiral, diastereoisomeric inclusion complexes are formed.

Sybilska and associates[29] have used β-cyclodextrin as a chiral mobile-phase additive in the resolution of mephenytoin, methylphenobarbital, and hexobarbital. They attribute the observed resolution to two different mechanisms. The resolution of mephenytoin is due to a difference in the absorption of the diastereoisomeric complexes on the achiral C-18 support. For methylphenobarbital and hexobarbital, the relative stabilities of the diastereoisomeric complexes are responsible for the resolution of these compounds.

In addition to the compounds listed here, mobile phases modified with β-cyclodextrin can resolve mandelic acid and some of its derivatives.[30, 31] Aqueous mobile phases modified with a buffer such as sodium acetate are commonly used. Alcoholic modifiers, such as ethanol, can be added to the mobile phase to reduce retention. Automation is possible for the direct measurement of biological samples. Sybilska and associates[29] have used it for preparative separations of mephenytoin, methylphenobarbital, and hexobarbital.

The applications of this approach seem limited. For example, unlike hexobarbital and methylphenobarbital, the chiral barbiturates secobarbital, pentobarbital, and thiopental are not resolved when chromatographed with a β-cyclodextrin-containing mobile phase.[27] Further discussion on solutions involving cyclodextrin inclusion complexes is included in the next section.

Enantiomeric Resolution using Chiral Stationary Phases

Enantiomers can be resolved by the formation of diastereomeric complexes between the solute and a chiral molecule that is bound to the stationary phase. The development of these chiral stationary phases is the fastest-growing area of chiral separations. The first commercially available HPLC-CSP was developed by Pirkle in 1981.[32] Currently, a large number of chiral phases are commercially available.

The separation of enantiomeric compounds on a CSP can be accomplished as a result of differences in energy between temporary diastereomeric complexes formed between the solute isomers and the CSP; the larger the difference, the greater the separation. The observed retention and efficiency of a CSP is the total of all the interactions between the solutes and the CSP, including achiral interactions.

Selection Procedure for CSPs

A classical classification of chiral stationary phases in five types is provided in Table 8.1. The table also indicates the usual mode of operation of these CSPs. A flow diagram for the selection of optimum CSP for a given sample is shown in Figure 8.1. The chromatographic analysis of enantiomeric compounds can also be broadly reviewed as normal-phase or reversed-phase type of separations.[33]

Normal-phase analyses. Type 1 or 2 CSPs are preferable. Derivatization by an achiral reagent can be used to increase analyte-CSP interactions and hence improve resolution of enantiomers. It can also bring other benefits, such as increased detectability by UV or fluorescence detectors, or enhanced solubility in normal-phase eluents. If derivatization is possible, a type 1 CSP (donor-acceptor) is favored. Otherwise, a type 2 CSP (silica-supported cellulose triesters, etc.) is selected except when a π-acid or a π-base is already located close to a chiral center in the analyte. For example, an arylsulfoxide is better separated on a type 1 CSP.

Table 8.1 Classification of Chiral Stationary Phases

Type	Description	Examples	Usual Mode
1	Brush-type (donor acceptors)	DNB-glycine DNB-leucine, naphthylalanine	Normal phase (polar modifier)
2	Cellulose triesters or carbamates on silica	Chiralcel OA, OB, OF, OJ, etc	Normal phase (polar modifier)
3	Inclusion CSPs, e.g., cyclodextrins, crown ethers, polyacrylates, polyacrylamides	Cyclobond 1-3, Chiralcel CR, Chiralpak OP,OT, Merck grafted	Reversed-phase, e.g., aqueous MeCN or aqueous MeOH
4	Ligand exchangers	Proline, OH-proline	Reversed-phase (aqueous buffers)
5	Proteins	Albumin, glycoprotein	Reversed-phase (aqueous buffers)

The analytes identified as candidates for separation with a type 1 CSP can be further divided into those with π-donor groups or π-acceptor groups. When there is a π-donor group (i.e., phenyl or naphthyl) close to the asymmetric center, a π-acceptor type 1 CSP such as the dinitrobenzoyl (D)-phenyl-glycine phase may be initially selected. Otherwise a π-donor type 1 CSP (e.g., naphthylvaline) would be indicated if a π-acceptor group was similarly located in the analyte or its derivative. These requirements are not overriding, however, since type 1 CSPs also operate by dipole-dipole, H-bonding, and steric interactions.

Reversed-phase (RP) analyses. If bidentate ligation to a metal ion such as Cu(II) is possible, type 4 CSPs (ligand exchange) are indicated. In the absence of such potential ligation, the selection process centers around CSPs of types 3 and 5.

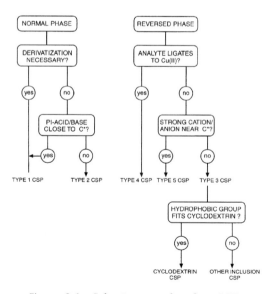

Figure 8.1 Selection procedure for a CSP.

For nonligating RP analytes, if a strongly cationic or anionic site is located close to the asymmetric center, the analyte may be resolvable on a protein-based CSP (type 5). Alternatively, inclusion CSPs (type 3) should be tried first.

Inclusion analyses. If a nonpolar includable group (ideally 5–9 Å across) is present (not too far from an asymmetric center), the selection of one of the Cyclobond range of cyclodextrin CSPs is indicated. The absence of a nonpolar includable group may lead to selection of other types of inclusion CSPs, such as the chiral acrylate polymers currently marketed by Merck and Daicel.

Macrocyclic antibiotics have been investigated as a new class of chiral selectors for liquid chromatography.[34] These stationary phases appear to be multimodal in that they can be used in both normal- and reversed-phase modes.

Since there are so many HPLC-CSPs available to the chromatographer, it is difficult to determine which is most suitable in solving a particular problem. This difficulty can be partially overcome by grouping the CSPs for chiral separations according to a common characteristic. The first step, that is, the formation of the solute-CSP complexes, is more readily adaptable to the development of a classification system.

Using this criterion for the division of CSP groups, the current commercially available CSPs can be divided into five categories.[33] Let's review the various types of stationary phases to develop a better understanding of them.

Pirkle-type and related phases (Type 1). Pirkle-type and related phases are often based on an amino acid derivative. The columns are available on π-acid or π-base columns (amide, urea, or ester moiety in CSP). They show the following solute-stationary phase interactions: dipole-dipole, hydrogen bonding, and π-π interactions. Polar functional groups—for example, amino, carboxyl, and some hydroxyl groups—require derivatization to less polar derivatives (Table 8.2). Recently developed hybrid columns can resolve a variety of underivatized compounds.[35]

MOBILE-PHASE SELECTION. The mobile phase usually contains a nonpolar organic solvent with varying amounts of a polar modifier—hexane/isopropyl alcohol, for example. Dichloromethane and ethanol have been used also. The mobile

Table 8.2 Derivatizations with Achiral Reagents

Group	Reagent	Product
NH	COOH derivative	Amide
	Isocyanate	Urea
	Chloroformate	Carbamate
COOH	Alcohol	Ester
	Aniline	Anilide
	Amine	Amide
OH	COOH derivative	Ester
	Isocyanate	Carbamate

phases are selected on the basis of their ability to optimize polar attractive interactions between the solute and the CSP.

Several studies have dealt with the structure and the concentration of the polar mobile-phase modifier on retention and selectivity.[35] The increase in the steric bulk of the modifier generally results in increase in the observed resolution of enantiomers—for example, t-butanol gave a higher α value than 2-propanol and ethanol. It has been theorized that increase in the bulk of alcohol enhances the ability of solute enantiomers to displace the modifier from the CSP that stabilizes the diastereomeric solute/CSP complexes. The enhanced stability of the two solute/CSP complexes magnifies the energy differences between them, leading to an increase in observed enantioselectivity.

Aqueous methanol mobile phases containing phosphate buffer have also been found useful for the resolution of N-(3,5-dinitrobenzoyl) derivatives of α-amino acids. Similarly, ammonium acetate has been used in an ethanol:dichloromethane mobile phase to resolve a number of underivatized β-blockers on α-Burke 1 CSP. The organic components in the above-mentioned mobile phase can be replaced by acetonitrile for the N-(3,5-dintrobenzoyl)-substituted amino acids on an (R)-1-naphthylurea CSP.

Target design studies led to the development of Whelk-O 1 CSP, which incorporates some features important for the enantioselective recognition of naproxen—face-to-edge π-π interactions in addition to hydrogen bonding and face-to-face π-π interactions. This CSP provides great selectivity for naproxen enantiomers (α = 2.25) and related NSAIDs. Furthermore, it provides capability to resolve enantiomers of racemates from a host of functional classes.

Derivatized cellulose and related CSPs (Type 2).

A. *Derivatized Cellulose.* Cellulose is a crystalline polymer composed of β-D-glucose units. Crystalline cellulose is unable to withstand HPLC pressures, and therefore it has to be modified. Several approaches have been reported, which are based on derivatizing the available hydroxyl groups, cellulose triacetate (CTA) and cellulose tribenzoate (CTB).

MOBILE-PHASE SELECTION. Largely aqueous mobile phases cannot be used with CTA-type CSPs. Alcohols such as ethanol are commonly used. Ethanol can be used in its anhydrous form or, at most, with 5% water. Alcohols such as methanol and 2-propanol can also be used and show dramatic effects. A number of studies led to the conclusion that the steric bulk rather than the polarity of eluent may be the key to their eluotropic strength. The alcohols can affect the steric environment of the channels and cavities by binding to sites near or at the site of inclusion, and thus change the fit of the solutes in the cavities. This affects both α and k' values. Nonpolar mobile phases composed of hexane modified with 2-propanol have also been used for CTA; however, they are more commonly used for CTB.

A variety of analytes can be resolved. CTA prefers the phenyl group. It is also well suited for preparative work. Multiple interactions are involved in chiral recognition, which includes inclusion into channels and cavities.

Table 8.3 Derivatized Cellulose Columns

Name	Separation	Mobile Phase
Chiralpak OT (+) Chiralpak OP (+)	Compounds with aromatic group	Methanol or hexane-2-propanol
Chiralcel OA Chiralcel OB Chiralcel OC Chiralcel OK Chiralcel OD	Compounds with aromatic group, carbonyl group, nitro group, sulfinyl group, cyano group, and hydroxyl group	Hexane-2-propanol or ethanol + water
Chiralcel CA1		Ethanol + water
Chiralpak WH Chiralpak WM Chiralpak WE	DL Amino acid or its derivative	Aqueous $CuSO_4$

B. *Derivatized Cellulose on Silica Gel CSPs.* Cellulose is depolymerized to shorter units, then derivatized to an ester or carbamate derivative, followed by coating onto silica, a noncovalent attachment (Table 8.3). A reversed-phase version of the OD CSP has been also marketed. In an OB CSP, at least two chiral recognition mechanisms have been shown to be operational: attractive interactions and hydrophobic insertion.

MOBILE-PHASE SELECTION. The mobile phases usually contain hexane-isopropyl alcohol and base or acid additive, depending on the analyte. The chiral recognition mechanism is not well known; multiple mechanisms, including fit into cavities, are possible. These mobile phases are used to maximize the attractive interactions between the solute and CSP. Since these interactions have a polar component, an increase in the polarity of the mobile phase with a larger alcohol content reduces the retention, and vice versa. Though some compounds require derivatization, many can be resolved without it.

Cyclodextrins and other inclusion-complex-based CSPs (Type 3).

Cyclodextrins are cyclic oligosaccharides of α-D-glucose formed by action of *Bacillus macerans* amylase on starch. Six, seven, or eight glucose units are possible, in α-, β-, and γ-cyclodextrin, respectively. They have toroidal form, and the interior is relatively hydrophobic. The size of the interior cavity varies: $\alpha < \beta < \gamma$. The naphthyl group can fit into a β-cyclodextrin cavity. Chiral recognition is based on entry of the solute or a portion of the solute molecule into the cavity to form an inclusion complex stabilized by hydrogen bonds and other forces.

MOBILE PHASE SELECTION. The mobile phases are generally composed of aqueous (often buffered) methanol, ethanol, or acetonitrile mixtures. Buffers such as triethylammonium acetate and phosphates can be used and tend to improve the efficiency and reduce retention of cationic and anionic solutes. As a general rule, the increase in the aqueous content of the mobile phase increases the retention and selectivity of these phases. Polar organic acetonitrile (ACN) mobile phase (with TEA/HOAc) has been used recently.

Several CSPs are obtained by derivatizing OH groups, to produce acetylates or other esters and carbamate derivatives—for example, N-1-(l-naphthyl)ethylcarbamate derivatives (R, S, RS). The chiral recognition mechanism does not appear to be an inclusion complex. These can be used with organic mobile phases, such as hexane/isopropyl alcohol.

Ligand exchangers (Type 4). As discussed, the technique of ligand exchangers is based on complexation to metal ion—Cu(II), for example. The chiral ligand immobilized on the column is usually an amino acid such as proline. Solutes must be able to form coordination complexes with a metal ion. Examples are amino acids and derivatives, hydantoins, and amino alcohols.

MOBILE-PHASE SELECTION. The mobile phase contains the metal ion, typically added as copper sulfate. The common concentration is 0.25 mM of copper(II) sulfate; however, concentrations as low as 0.05 mM and as high as 20.0 mM have been used. Retention and resolution can be manipulated with concentration of the metal ion, pH, addition of modifiers, and temperature.

Protein CSPs (Type 5). Proteins are polymers composed of chiral units (L-amino acids) and are known to be able to bind small organic molecules. The protein can be immobilized on solid support by a variety of bonding chemistries; the choice of bonding method affects selectivity of the CSP obtained. The following protein CSPs have been used: α acid glycoprotein (AGP), ovomucoid (a glycoprotein from egg white), human serum albumin, and bovine serum albumin.

Retention and chiral recognition mechanisms may be unrelated to drug-protein binding by free protein (i.e., in vivo or in vitro). Retention is most likely based on hydrogen bonding, ionic interactions, and so on. Recent investigations with commercial chicken ovomucoid indicate that chiral recognition ability may be due to an ovoglycoprotein which is present as an impurity.[36]

MOBILE-PHASE SELECTION. The mobile phases are generally aqueous buffers within a limited pH range with a variety of organic modifiers. They can be charged or uncharged compounds, and include 2-propanol, octanoic acid, N,N-dimethyldioctyl amine, and tetrabutylammonium bromide. The addition of these modifiers can show improvement in α values with a decrease in retention values. To manipulate retention and stereoselectivity, one can vary mobile-phase composition, pH, and temperature. The capacity of protein CSPs is limited; however, they enjoy broad applicability.

Computerized Methods

These methods are based on evaluation of the potential energy contours obtained in molecular docking processes. Thus, by computerized matching of the contours found from the docking of two enantiomeric molecules with the same chiral entity, an energy difference between the two complexes at their potential energy minima can be calculated. This offers, at least in principle, a means of predicting the elution order in a chromatographic system. However, solvent effects cannot be taken specifically into account by such methods, and, so far, the predictive value is rather limited. Furthermore, even in very well-defined chromatographic systems containing a simple chiral selector, different retention mechanisms may compete, making the situation more complex than assumed in the theoretical approach. Nevertheless, data from computational work on chiral discrimination will continue to add to our understanding of molecular interactions.[37, 38]

Chirule, an aid in developing chiral separations, has been described by Stauffer and Dessy.[39] It constructs an N-dimensional information space from a large number of known chiral separations by fragmenting the molecules at their chiral center.

REFERENCES

1. S. Ahuja, *Chiral Separations: Applications and Technology*. American Chemical Society: Washington, DC, 1997.
2. S. Allenmark, *Chromatographic Enantioseparation*. Ellis Horwood: New York, 1991
3. S. Ahuja, *Chiral Separations by Liquid Chromatography*, ACS Symposium Series #471. American Chemical Society: Washington, DC, 1991.
4. S. Ahuja, *Selectivity and Detectability Optimizations in HPLC*. Wiley: New York, 1989
5. W. J. Lough, *Chiral Liquid Chromatography*. Blackie and Son: Glasgow, UK, 1989.
6. J. Gal, *LC-GC*, **5**, 106 (1987).
7. S. Hara and J. Cazes, *J. Liq. Chromatogr.*, **9**, Nos. 2 and 3 (1986).
8. R. W. Souter, *Chromatographic Separations of Stereoisomers*. CRC Press: Boca Raton, FL, 1985.
9. S. Ahuja, First International Symposium on Separation of Chiral Molecules, Paris, May 31–June 2, 1988.
10. G. Blaschke, H. P. Kraft, K. Fickentscher, and F. Koehler, *Arzneim.-Forsch*, **29**, 1640 (1979).
11. S. Ahuja, *The Impact of Stereochemistry on Drug Development and Use*, H. Y. Aboul-Enein and I. W. Wainer, Eds. Wiley: New York, 1997, p. 287.
12. Y. Okamoto, *CHEMTECH*, 176 (March 1987).
13. W. H. DeCamp, *Chirality*, **1**, 2 (1989).
14. *Guidelines for Submitting Supporting Documentation in Drug Applications for the Manufacture of Drug Substances*. Office of Drug Evaluation and Research (HFD-l00) Food and Drug Administration: Rockville, MD, 1987.
15. S. Ahuja, *Chromatography of Pharmaceuticals*, ACS Symposium Series #512, American Chemical Society: Washington, DC, 1992.
16. M. D. Palamareva and L. R. Snyder, *Chromatographia*, **19**, 352 (1984).
17. J. Bitterova, L. Soltes, and T. R. Fanovee, *Pharmazie*, **45**, H.6 (1990).
18. E. Gil-Av In *Chiral Separations by Liquid Chromatography*, ACS Symposium Series #471, S. Ahuja, Ed. American Chemical Society: Washington, DC, 1991.
19. V. Schurig, M. Jung, D. Schmalzing, M. Schliemer, J. Duvekot, J. C. Buyten, J. A. Peene, and P. Musschee, *J. High Res. Chromatogr.* **13**, 470 (1990).
20. C. Dalgliesh, *J. Chem. Soc.*, 137, 1952.
21. G. Wulff, in *Polymeric Reagents and Catalysts*, W. T. Ford, Ed. American Chemical Society: Washington, DC, 1986, p. 186.
22. I. Wainer and D. E. Drayer, Eds., *Drug Stereochemistry, Analytical Methods and Pharmacology*. Marcel Dekker: New York, 1988.
23. R. Dappen, H. Arm, and V. R. Mayer, *J. Chromatogr.*, **373**, 1 (1986).

24. B. Silber and S. Riegelman, *J. Pharmacol. Exp. Ther.*, **215**, 643 (1980).
25. I. S. Krull, in *Advances in Chromatography*, Vol. 16, J. C. Giddings, E. Grushka, J. Cazes, and P. R. Brown, Eds. Lippincott: Philadelphia, PA, 1977, p. 146.
26. W. H. Lindner and C. Pettersson, in *Liquid Chromatography in Pharmaceutical Development*, I. W. Wainer. Ed. Aster: Springfield, MA, 1985, p. 63.
27. C. Pettersson and G. Schill, *J. Liq. Chromatogr.*, **9**, 269 (1986).
28. S. Allenmark, *J. Liq. Chromatogr.* **9**, 425 (1986).
29. D. Sybilska, J. Zukowski, and J. Bojarski, *J. Liq. Chromatogr.*, **9**, 591 (1986).
30. J. Debowski, D. Sybilska, and J. Jirczak, *J. Chromatogr.*, **237**, 303 (1982).
31. J. Debowski, D. Sybilska, and J. Jirczak, *J. Chromatogr.*, **282**, 83 (1983).
32. W. H. Pirkle, J. M. Finn, J. L. Schreiner, and B. C. Hamper, *J. Am. Chem. Soc.*, **103**, 3964 (1981)
33. D. R. Taylor, in *Recent Advances in Chiral Separations*, D. Stevenson and I. D. Wilson, Eds. Plenum: New York, 1990.
34. D. W. Armstrong, Y. Tang, S. Chen, Y. Zhou, C. Bagwill, and J-R. Chen, *Anal. Chem.*, **66**, 1473 (1994).
35. C. J. Welch, in *Advances in Chromatography*, P. R. Brown and E. Grushka, Eds. Marcel Dekker: New York, 1995, p. 171.
36. J. Haginaka, C. Seyama, and N. Kanasugi, *Anal. Chem.*, **67**, 2539 (1995).
37. K. B. Lipkowitz and B. Baker, *Anal. Chem.*, **62**, 774 (1990).
38. S. Topol, *Chirality*, **1**, 69 (1989).
39. S. T. Stauffer and R. E. Dessy, *J. Chromatogr. Sci.*, **32**, 228 (1994).

REVIEW

1. Suggest a CSP for each of the following compounds:

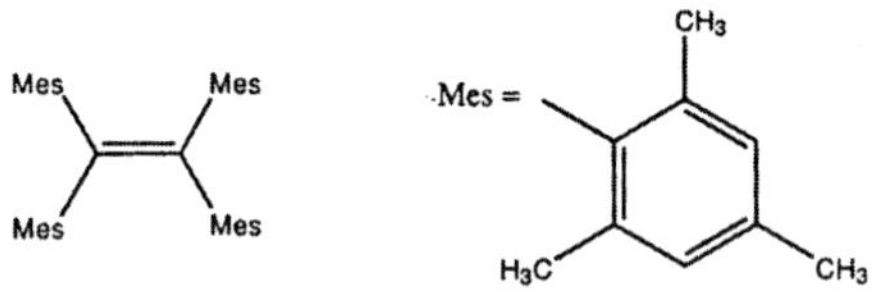

tetramesitylethylene

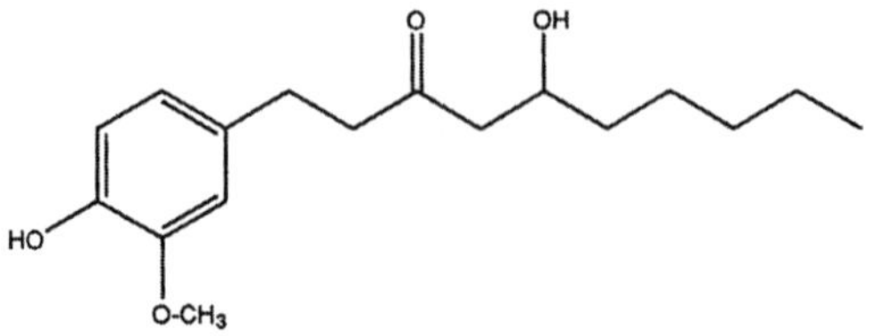

gingerol

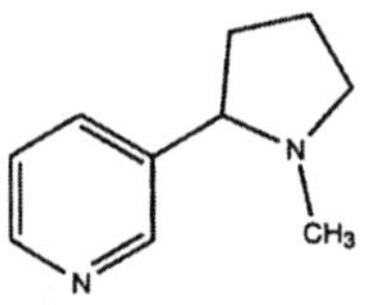

nicotine

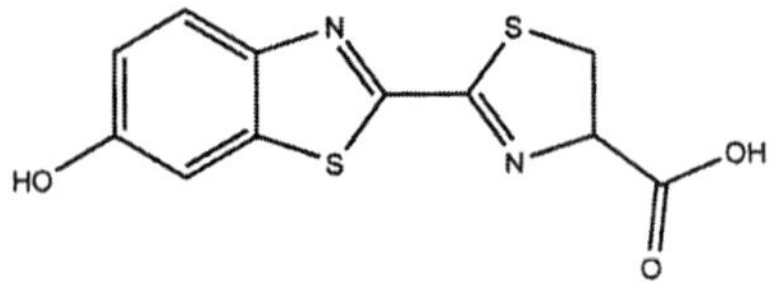

luciferin

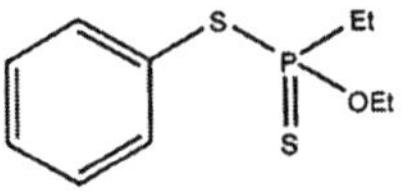

fonofos

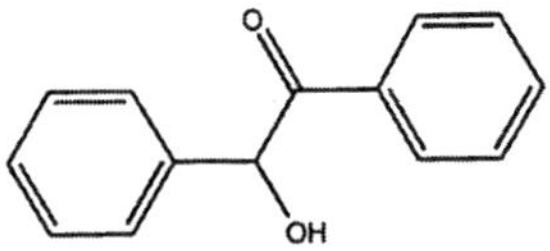

benzoin

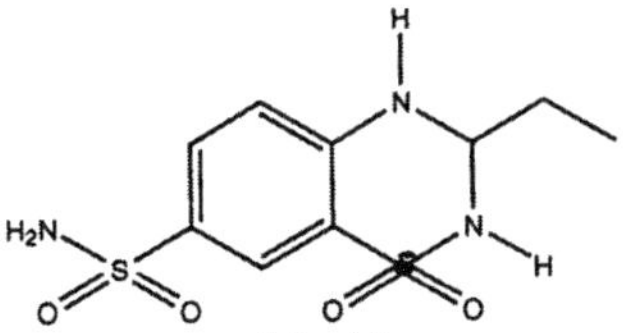

ethiazide

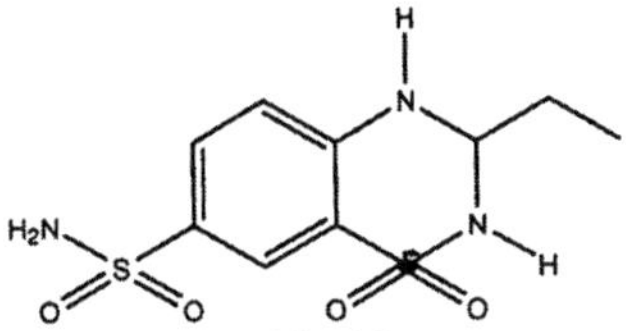

ethiazide

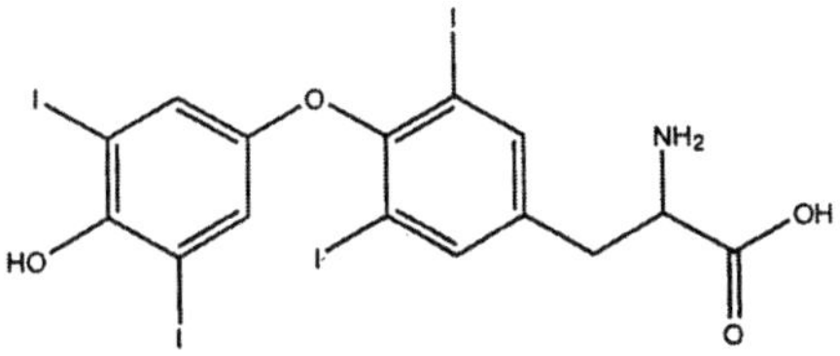

thyroxine

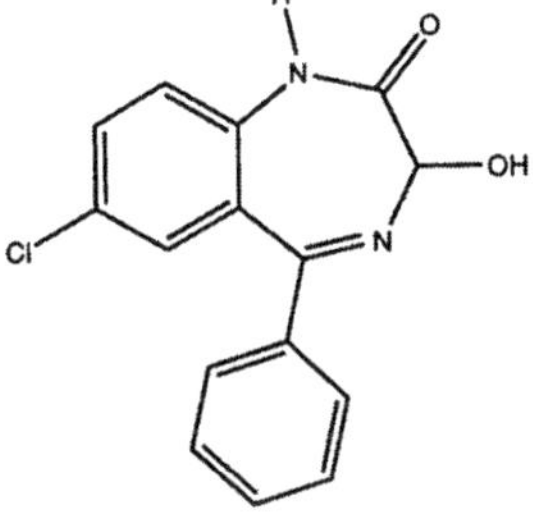

oxazepam

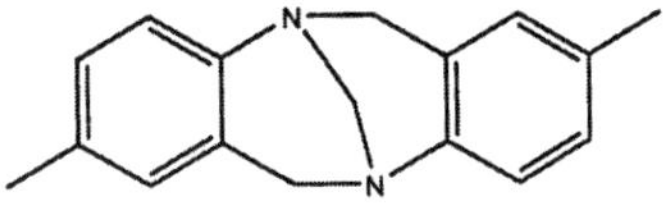

Troeger's Base

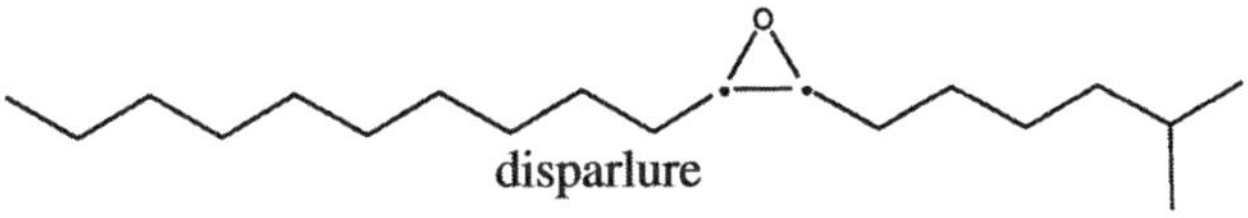

disparlure

TUTORIAL

To choose the correct CSP:

1. Do literature and database search.
   a. Effective use of the literature can greatly simplify the task of developing an analytical method.
   b. Specific databases for chromatographic chiral separations are commercially available: CHIRBASE GC or LC.

This database is expensive; however, it is quite useful. For more information, contact the U.S. Distributor:

Separacorp, Inc.
33 Locke Drive
Marlborough, MA 01752
tel. (508)481-6700
FAX (508) 481-7683

Some tips for a CAS search:

- Combined searches by compound name or CAS registry number with the word "enantiomer" are usually successful.
- Search for methods developed for similar compounds or general compound classes.
- Consult standard texts on chromatographic enantioseparations.

2. Perform conformational analysis. It is often useful to consider the preferred conformation of analyte molecule. Low-energy conformations as determined by computer programs (e.g., Chem 3-D) are often useful, although these minimizations typically do not take account of solvent or many weaker interactions (e.g., face-edge π-π interaction, H-bonding to π-cloud, etc.).
   - Is one conformation strongly preferred?
   - If more than one conformation is possible, which would be preferred for chiral recognition?
   - Can equilibrium be shifted so as to form more heavily populated desired conformer?

   mobile-phase additives
   solvents that will disrupt or encourage intramolecular hydrogen bonding.

A number of books have been published on conformational analysis (e.g., see E. L. Eliel and S. H. Wilen, *Stereochemistry of Organic Compounds*. Wiley: NY. 1994. Most general organic textbooks have a treatment of this topic.

3. Make and use molecular models!
   - FMO or framework (Dreiding, Darling)
   - CPK or space-filling

9

Preparative Separations

There are two possible approaches for preparing commercially pure enantiomeric compounds.[1-3] The first is to design an elaborate stereoselective synthesis that would lead to production of pure enantiomers. The second consists of resolving racemic compounds into enantiomers; this usually leads to the production of both enantiomers.

Even though chromatography may appear to be an expensive technique to some scientists, it is increasingly regarded as both technically and economically attractive for preparation of high-value compounds and for optical isomers that are otherwise accessible only with great difficulty.

Preparative chromatography of enantiomers is generally performed for one or more of the following reasons:

- To evaluate biological properties of the desired compound
- To study chiroptical properties of the desired compound and intermediates
- To control synthesis of chiral intermediates to assure production of the desired intermediates
- To control the desired compound in terms of its enantiomeric purity

For therapeutic, toxicologic, and regulatory reasons, pharmaceutical compounds have been largely subjected to preparative separations to investigate their chiroptical and biological properties. It is frequently necessary to have a small amount of a given enantiomer for these studies. This is where preparative separations excel. Of necessity, it is important to control chiral intermediate synthesis to assure that the final product meets the necessary enantiomeric purity. Again, preparative separations may be necessary here to assure that the correct intermediates are being synthesized. Let's study various pathways to the development of a new chiral entity that may become a potential drug product (Figure 9.1).

Preparative-scale separations can be carried out by any of the following techniques: frontal chromatography, displacement chromatography, elution chromatography.

In the frontal chromatography technique, the sample is fed continuously into the chromatographic column. This produces a series of concentration steps (fronts), each corresponding to a separate component of the sample. The fastest-moving component is physically separated from the other substances. Generally this separation is partial, at best. Therefore, the other two techniques are more commonly used.

Displacement chromatography is an elution method that requires the mobile phase to contain a component or components generally referred to as a displacer that is more strongly retained by the stationary phase than the sample components being studied. The latter are then forced out, or displaced, from the stationary phase and subsequently eluted.

Elution chromatography is a simple elution method commonly used by most chromatographers. It entails introducing a small amount of the sample to be analyzed into the flowing mobile phase (eluent) and observing the separation of the various components of the sample after their passage through the chromatographic column, in the form of concentration bands or peaks.

Because of their relative importance in preparative separations, displacement chromatography and elution chromatography are discussed further in the next sections.

Displacement Chromatography

The following important requirements have to be met to achieve a successful separation by displacement chromatography:

1. The sample should be retained from the carrier solvent—that is, $k' > 1$.
2. The displacer should be more highly retained than the sample.
3. The isotherms for the sample and displacer should be concave.
4. The sample and displacer should be highly soluble in the carrier solvent.

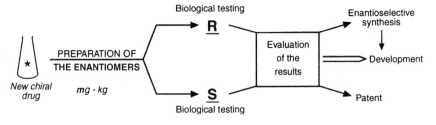

Figure 9.1 Pathways to an enantiomer.

The displacement process can be diagramatically represented as shown in Figure 9.2 for three sample components 1, 2, and 3. For the displacement to occur successfully, k carrier << k1 < k2 << k3 < k displacer. Displacement chromatography offers the following advantages: high efficiency due to self-focusing mechanism, high product concentrations, and high column loadings. If the right type of displacer is found and the sample to be separated meets the criteria, displacement is a favored technique. However, the selection of a suitable displacer can be quite difficult. This tends to make elution chromatography more favorable.

Elution Chromatography

Most preparative separations are developed after a successful analytical separation has been achieved. Since most chromatographic analytical separations are based on elution chromatography, this technique is generally favored over all other techniques for preparative separations. Therefore a significant effort must be applied in designing the scale-up of analytical separations.

Scale-up of Analytical Separations

For scale-up of an analytical method, the following considerations are applied:

1. Determine the capacity of the CSP.
2. Consider the solubility of the compounds in the mobile phase.

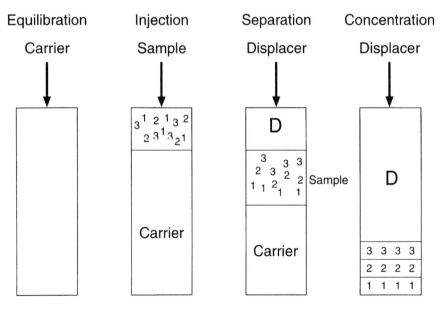

Figure 9.2   A diagrammatic representation of the displacement process.

3. Avoid injection of the sample in a nonmobile-phase solvent.
4. Evaluate various resolution strategies.
5. Determine which method provides maximum material in the minimum time.
6. Consider automation and solvent recovery.

Discussed next are some of the important considerations that need to be reviewed with some care before proceeding with scale-up.

Capacity of the CSP

The capacity of the CSP in a given column size can be determined by making several injections at various concentrations. This is exemplified by a number of chromatograms shown in Figure 9.3, where samples of various concentrations of abscisic acid were injected onto an (S,S) Whelk O-1 (4.6 mm i.d. × 25 cm length) column. The results show that this column is definitely overloaded at 6.2 mg per injection level, so preparative separations are best carried out at around 3 mg or a little higher level.

Solubility of the Sample

It is important to evaluate the solubility of a sample in the mobile phase in HPLC. High solubility in the mobile phase suggests that relatively large amounts of the compound of interest can be injected in a minimal volume. Larger injection volumes for poorly soluble samples can lead to broad peaks. It may be preferable to change the mobile phase to improve the solubility of the sample in case comparable resolution can still be obtained. Injections in a solvent different from the mobile phase can lead to problems in that the sample could precipitate at the head of the column. This can lead to a host of chromatographic problems, such as severe peak tailing, peak doubling, or total loss of resolution. Mechanical problems such as plugged flow, high back pressure, blown seals, clogged injectors, and so forth can also arise.

If it is absolutely necessary to inject in a different solvent, try to find a solvent that does not lead to sample precipitation when injected onto a column. 1,4-Dioxane often works well in this regard.

Resolution Strategy

Broad peaks allow high production if high enantiomeric purity is not necessary or when it can be combined with a cleanup step. Narrow peaks are desirable when only a small number of enantiomers are needed and further manipulations are considered excessively time-consuming. It is a good idea to opt for a method that provides maximum material in minimum time. The peak capacity of the first peak should generally be higher than 1 and lower than 5 for preparative separations.

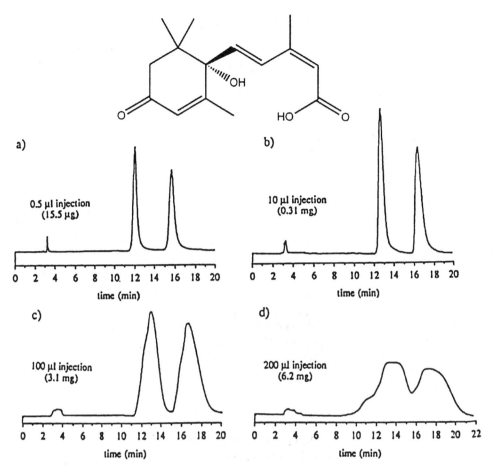

Figure 9.3 Semipreparative separations of enantiomers of abscisic acid. Reproduced from S. Ahuja, W. Pirkle, and C. Welch, *Chiral Separations by Chromatography*, American Chemical Society Short Course, 1994. Copyright 1994 American Chemical Society.

Automation and Solvent Recovery

Do automate preparative separations if possible, as a great deal of tedious work of repeated injections and collection of fractions can be eliminated. Automation should include at least sample injection and collection of the appropriate fractions.

Preparative chromatography can use a large amount of solvent that can result in high initial solvent costs, high solvent-disposal costs, concerns about health and safety hazards, and investment in equipment for solvent evaporation or recovery. Some of these problems can be solved by using in-line distillation or recycling the mobile phase.

Chiral Stationary Phases (CSPs)

The chiral phases used for preparative separations can be classified as follows:

1. Naturally occurring polymers and their derivatives

   Cellulose, dextran, and starch
   Cellulose triacetate (I & II)
   Benzoylcellulose derivatives
   Cellulose carbamate derivatives

2. Synthetic chiral polymers

   Poly(meth)acrylamides
   Cross-linked cyclodextrins

3. Chiral-modified silica gels

   Cyclodextrin
   π Acid and π base

Ligand exchange can be carried out on CSPs listed under numbers 2 or 3 above. Figure 9.4 provides a flow diagram for performance of preparative separations starting with small-scale separations on an analytical column.

Some of the desirable features of preparative CSPs include the following: wide availability, easy preparation of the phase, relatively low preparation costs, durability (chemical and mechanical stability), high loading capacity, and broad loading capacity.

Selection of an appropriate CSP
(Variation of the solute)
↓
Improvement of the separation/resolution factor by
optimization of the chromatographic parameters
(Mobile phase, temperature, additives, flow rate)
↓
Optimization of the chromatographic throughput
(Sample amount, column overloading)
↓
Choice of the column dimension
(First test and adjustment of the chromatographic conditions)
↓
Performance of the preparative separation

Figure 9.4 A flow diagram for performing preparative separations.

Resolution

To achieve desirable separations of the compound of interest, it is necessary to obtain a resolution value of 1.2 to 1.3. CSPs that have been commonly used for preparative separations are given in Table 9.1.

The most widely used derivative of cellulose for enantiomeric separation is cellulose triacetate. Except for microcrystalline cellulose triacetate (CTA-I), which is used most often in the reversed-phase mode, the cellulose- and amylose-based CSPs have been used under both normal and reversed-phase modes. Improvement

Table 9.1 CSPs Commonly Used for Preparative Chromatography of Chiral Compounds

Packing Name	Chiral Selector	Supplier
Cellulose and amylose derivatives		
Chiralcel OA	Cellulose triacetate	Daicel
Chiralcel OB	Cellulose tribenzoate	Daicel
Chiralcel OC	Cellulose tris(phenylcarbamate)	Daicel
Chiralcel OD	Cellulose tris(3,5-dimethylphenylcarbamate)	Daicel
Chiralcel OJ	Cellulose tris(4-methylbenzoate)	Daicel
Chiralcel OK	Cellulose tris(cinnamate)	Daicel
Chiralpak AD	Amylose tris(3,5-dimethylphenylcarbamate)	Daicel
CTA-I	Cellulose triacetate	Daicel, Merck
CTB	Cellulose tribenzoate	Riedel-de Haen
Cyclodextrin-based CSPs		
ChiraDex	β-Cyclodextrin	Merck
Cyclobond I	β-Cyclodextrin	Astec
Hyd-β-CD	Hydroxypropyl-β-cyclodextrin	Merck
π-acidic and π-basic CSPs		
DNBLeu	3,5-Dinitrobenzoylleucine	Regis
DNBPG-co	3,5-Dinitrobenzoylphenylglycine (covalent bonding)	Regis
DNBPG-io	3,5-Dinitrobenzoylphenylglycine (ionic bonding)	Regis
ChyRoSine-A	3,5-Dinitrobenzoyltyrosine butylamide	Sedere
NAP-AI	Naphthylalanine	Regis
Whelk-O 1	3,5-Dinitrobenzoyltetrahydrophenanthrene amine	Regis
Polyacrylamide CSP		
ChiraSpher	Poly[(S)-N-acryloylphenylalinine ethyl ester]	Merck
CSPs for ligand-exchange chromatography		
Chirosolve-pro	L- or D-proline bound to polyacrylamide	JPS Chimie
Chirosolve-phe	L- or D-phenylalanine bound to polyacrylamide	JPS Chimie
Chirosolve-val	L- or D-valine bound to polyacrylamide	JPS Chimie
Chirosolve-hypro	L- or D-hydroxyproline bound to polyacrylamide	JPS Chimie
Chirosolve-porretine	L- or D-porretine bound to polyacrylamide	JPS Chimie

Source: Reproduced from E. Francotte in *Chiral Separations: Applications and Technology*, S. Ahuja, Ed., American Chemical Society, Washington, D.C., 1997. Copyright 1997 American Chemical Society.

or resolution is generally achieved by varying the composition of the mobile phase (see chapter 8).

Protein-based CSPs do not usually exhibit desirable characteristics for preparative separations; however, they can be useful for resolution of racemic compounds when only small quantities are required.

The π-acid and π-base phases are quite popular because they offer the opportunity to reverse the order of eluted enantiomers, such that preparative separation can be carried out favorably. They are generally operated in the normal phase; however, reversed-phase separations are also possible.

Selection of a Suitable Mode of Chromatography

A variety of methods can be used for preparative chromatography (see the Applications section of this chapter). Most separations are performed by HPLC by a batch-mode process, using an appropriate CSP. Interest in simulated moving-bed technology is growing because it permits increased productivity with minimal use of solvents, thus decreasing the production costs.[4] However, significant investments are necessary to set up this technology, and it is also difficult to apply. As a result, it is likely to be suitable for only those compounds that have to be produced on a large scale.

A limited number of applications have been reported with supercritical fluid chromatography;[5] however, problems still exist relating to removal of mobile phase, especially where modifiers have been used with carbon dioxide.

Method Development

The general strategy for method development can be summed up as follows:

1. Select an analytical column for preliminary separation whose CSP can also be used in the preparatory mode.
2. Screen the sample on selected analytical columns, and optimize the various chromatographic parameters.
3. Adjust chromatographic conditions for the preparatory column to obtain resolutions comparable to the analytical column.
4. Automate the preparatory separation.

If no separation is obtained on chiral columns, achiral derivatization may be investigated (see chapters 4 and 8).

Points made about method development in chapter 8 are also applicable here. In short, improvement in separation on CSPs can be made by adjusting mobile-phase composition, flow rate, temperature, and determination of optimum loadability. In some cases, the inversion of elution order can occur when the amount of racemate injected is increased from analytical scale to preparative scale. This effect was observed with CTA-I,[6,7] suggesting complexity of these phases and incomplete knowledge on separation mechanisms. Inversion of elution order can also occur in changing the composition of the mobile phase and/or pH with cellulose-based CSPs.[8,9]

Applications

The load capacity of various columns from analytical to preparative scale is given in Table 9.2 to provide an idea of the size of columns and the sample size that can be used with them. A review of this table, coupled with some knowledge of the sample, will help determine the number of runs that would have to be made to get the desired amount of a given enantiomer.

Drugs. A number of drugs have been resolved on various chiral stationary phases, as shown in Table 9.3. The structure, sample size, and reference to the original paper is also given to enable readers to adapt a given method for their application.

An example of separation of a potential nootropic drug that used recycling and peak shaving has been discussed by Eric Francotte.[10] Separation of the enantiomers of the nootropic compound (2 g of CGS 16920) was achieved on a 45 × 5 cm TBC column with 100% methanol as the mobile phase. The methanol was used at a flow rate of 60 mL/minute, and the recycled portion is shown in Figure 9.5.

Pesticides and pheromones. A number of examples of applications of fungicides, herbicides, insecticides, and pheromones are given in Table 9.4. In most cases, CTA-I has been used as the CSP, and recycling has improved the throughput.[11]

Compounds with chiral heteroatoms. It is often quite difficult to obtain the optically pure form of compounds containing heteroatoms by conventional approaches. Again, CTA-I is a popular CSP for these separations (Table 9.5).

Chiral sulfur and phosphorous compounds include drugs such as ifosfamide (see Drugs in the Applications section above). The examples given earlier in this chapter attest to the great flexibility of chromatographic techniques.

Planar chiral compounds. Table 9.6 lists separations of planar chiral compounds. CTA-I has been found useful for most of the separations given in the table.

Helical chirality. Table 9.7 shows resolution of compounds with helical or propeller chirality. CTA-I is favored as the CSP for this group of compounds.

Table 9.2 Load Capacity of Columns

Column	i.d.(mm)	Sample (mg)
Analytical	1–5	0.2–2
Semipreparative	5–20	2–100
Preparative	20–100	100–10^4
Industrial	100–1000	>10^4

Table 9.3 Separations of Drugs with a Variety of Columns

Drug	Structure	Sample Size	Footnote No.
1. *Separations on CTA I Column*			
Ketamine	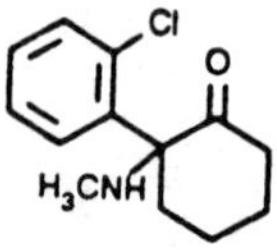	450 mg	1
Mianserin	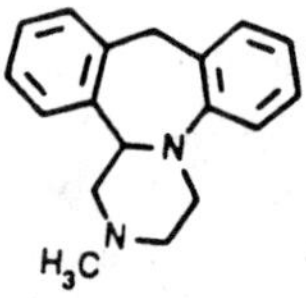	8 mg	1
Oxapadol	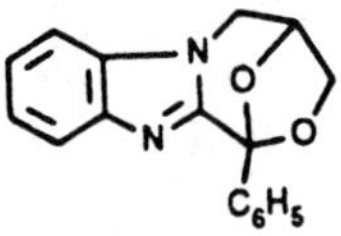	2.1 g	1
Rolipram	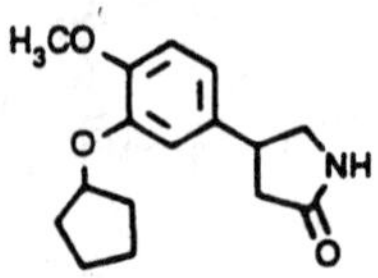	205 mg	1
Chlormezanone	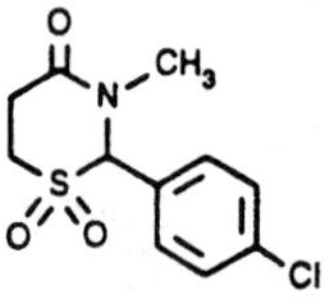	700 mg	2,3
Nefopam	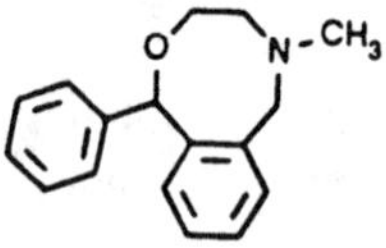	100 mg	4
Methaqualone	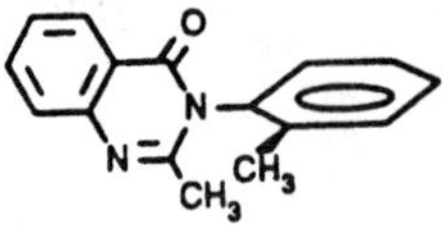	300 mg	5,6
Oxindanac	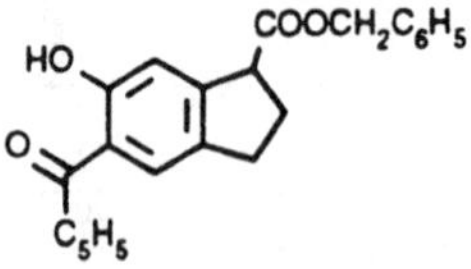	20 g	7,8

(*continued*)

Table 9.3 (Continued)

Drug	Structure	Sample Size	Footnote No.
Barbiturates	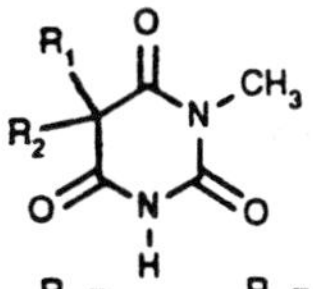 R₁ = Me, Et, n-Pr; R₂ = C₆H₅; Me, Et / cyclohexene; Me, Et / cyclohexane; Et / cyclopentene	257 mg	1,9
Praziquantel	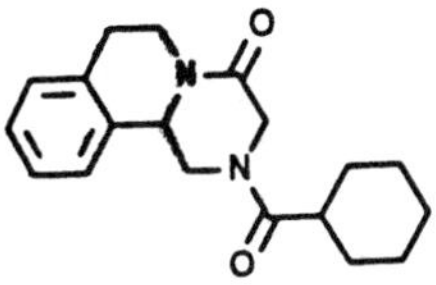	1g/hr	10
Etazepine acetate	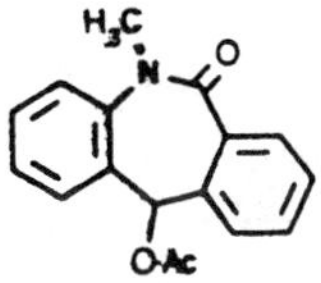	300 mg	11

2. *Separations on Poly-PEA Column*

Drug	Structure	Sample Size	Footnote No.
Chlorthalidone	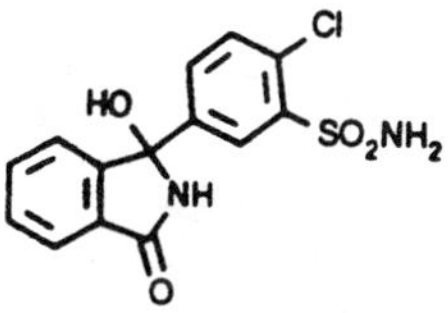	530 mg	1,12
Ifosfamide	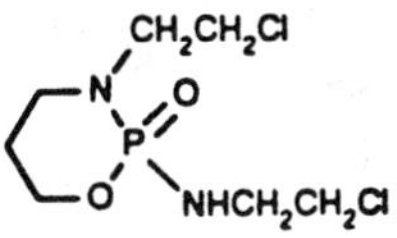	1.55 g	13,14
Penflutizide	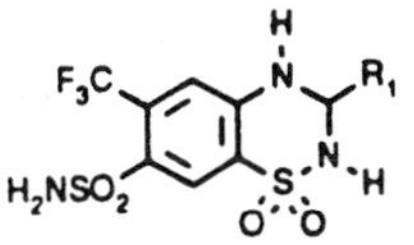 $R_1 = CH_2(CH_2)_4CH_3$	210 mg	15

(*continued*)

Table 9.3 (Continued)

Drug	Structure	Sample Size	Footnote No.
Bendroflumethiazide	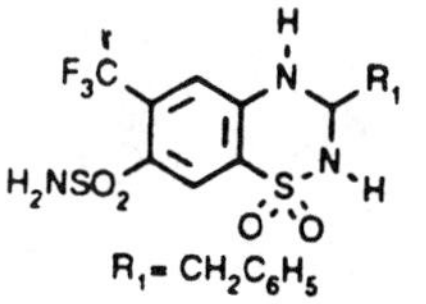 $R_1 = CH_2C_6H_5$	210 mg	15
Buthiazide	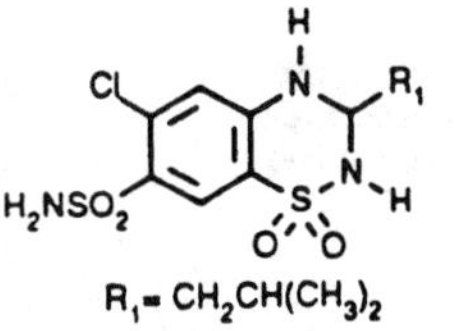 $R_1 = CH_2CH(CH_3)_2$	253 mg	15
Oxazepam	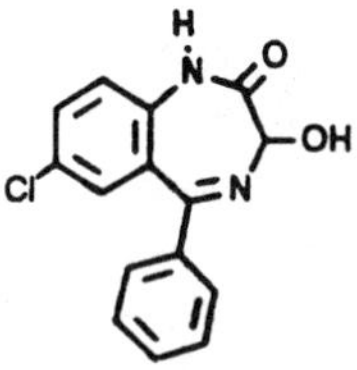	46 mg	12
Thalidomide	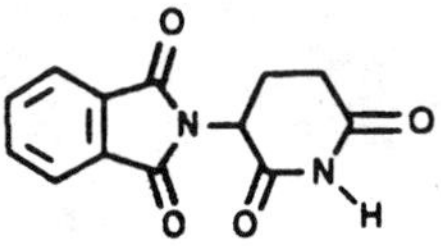	505 mg (poly-CHMA)	16,17

3. *Separations on Chiralcel OD Column*

Drug	Structure	Sample Size	Footnote No.
Propanolol	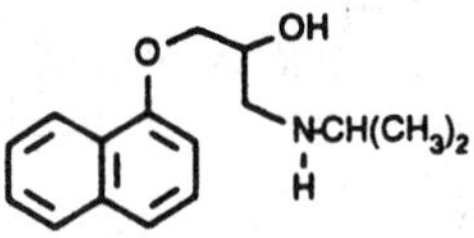	100 mg	18
Alprenolol	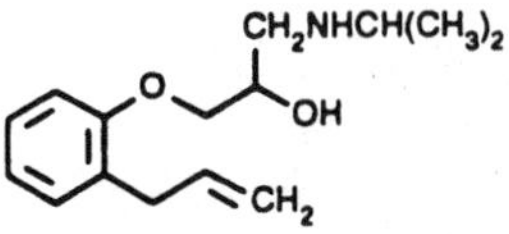	150 mg	18,19
Oxyprenolol	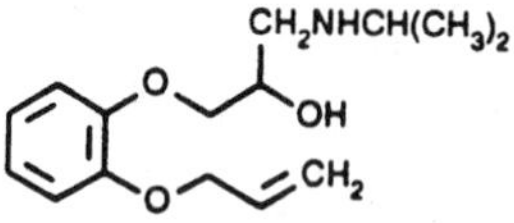	400 mg	18

(*continued*)

Table 9.3 (Continued)

Drug	Structure	Sample Size	Footnote No.
Ifsofamide	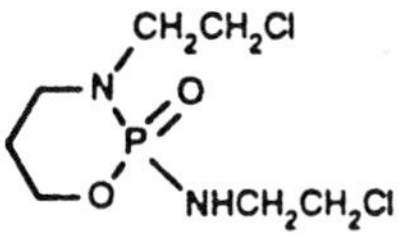	8 mg	20
Warfarin	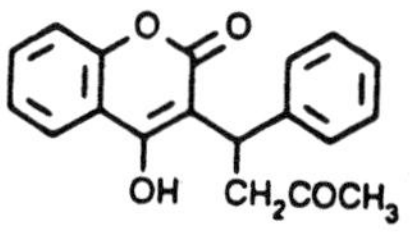	Not given	21
Lifibrol metabolite	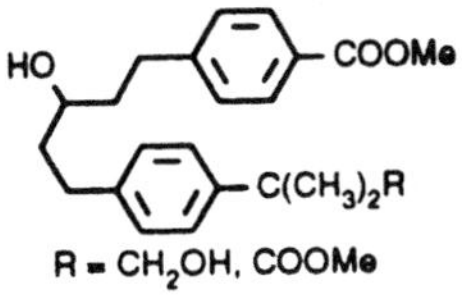	250 mg	22

4. Separations on Chiralcel OC Column

Omeperazole	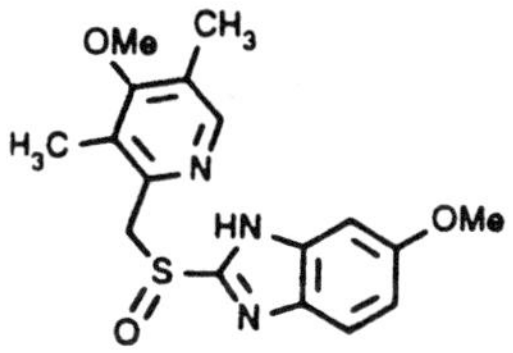	5 mg	23
Prostaglandin intermediate	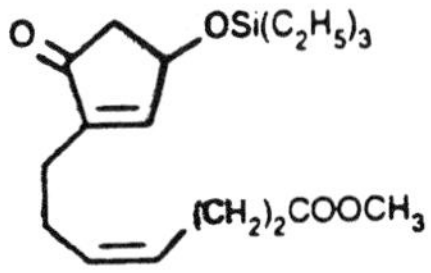	500 mg	24

5. Separations on DNBPG Derivative Column

Oxazepam	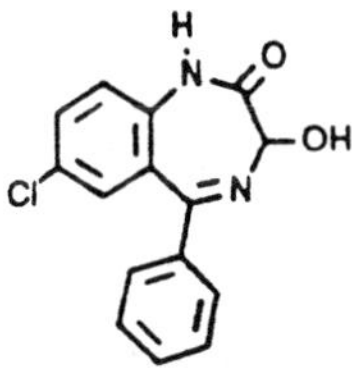	5 mg	25
Oxazepam pivalate		200 mg	26
Benzodiazepinones	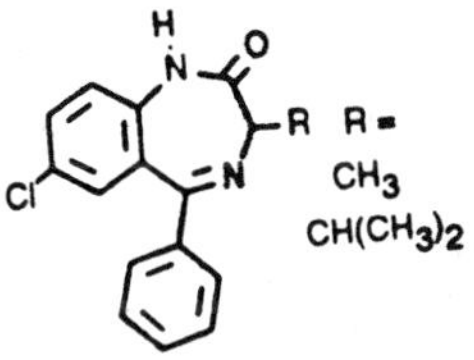	1 g	27

(continued)

Table 9.3 (Continued)

Drug	Structure	Sample Size	Footnote No.
Lorazepam	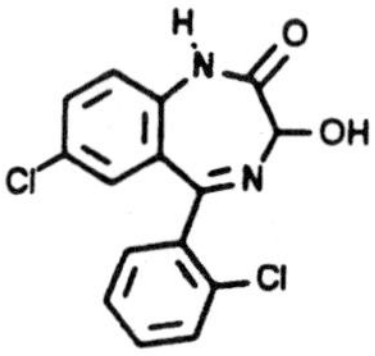	5 mg	25

6. Separations on Chiral AGP Column

Bendroflumethiazide	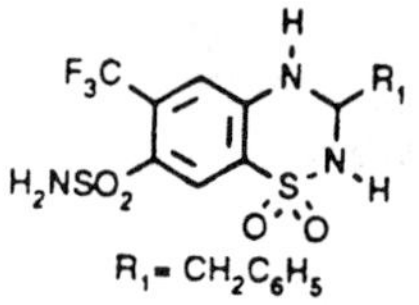 $R_1 = CH_2C_6H_5$	2.32 mg	28
Proglumide	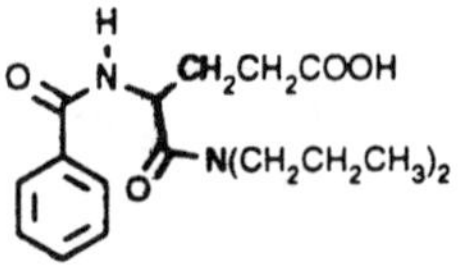	0.84 mg	28
Disopyramide	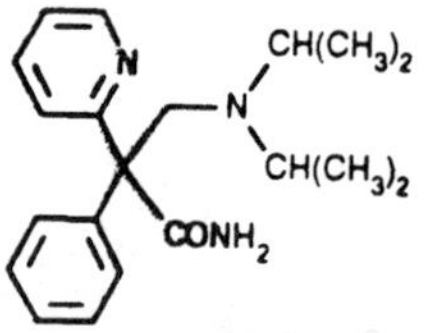	0.8 mg	28

1. G. Blaschke, *J. Liq. Chromatogr.*, **9**, 341 (1986).
2. S. Allenmark and R. A. Thomson, *Tetrahedron Lett.*, **28**, 3751 (1987).
3. G. Blaschke, W. Fraenkel, B. Frohlingsdorf, and A. Marx, *Liebigs Ann. Chem.*, 753 (1988).
4. R. Isaksson, J. Sandstrom, M. Elias, Z. Israely, and I. Agranat, *J. Pharm. Pharmacol.*, **40**, 48 (1988).
5. K. Rimbock, F. Kastner, and A. Mannschreck, *J. Chromatogr.*, **329**, 307 (1985).
6. A. Manschreck, H. Koller, G. Stuhler, M. Davies, and J. Traber, *Eur. J. Med. Chem. Chim. Ther.*, **19**, 381 (1984).
7. E. Francotte, H. Stierlin, and J. Faigle, *J. Chromatogr.*, **346**, 321 (1985).
8. D. Lohman, K. Auer, and E. Francotte, International Symposium on Chromatography in Israel, 1988.
9. G. Blaschke and H. Markgraf, *Arch. Pharm.*, **317**, 465 (1984).
10. C. Ching, B. Lim, E. Lee, and S. Ng, *J. Chromatogr.*, **634**, 215 91993).
11. E. Francotte, unpublished, through ref. 2 of this chapter.
12. G. Blaschke and H. Markgraf, *Chem. Ber.*, **113**, 2031 (1988).
13. G. Blaschke, P. Hilgard, J. Maibaum, U. Neimeyer, and J. Pohl, *Arzneim. Forsch.*, **36**, 1493 (1986).
14. G. Blaschke and J. Maibaum, *J. Chromatogr.*, **366**, 329 (1986).

Table 9.3 (*Continued*)

15. G. Blaschke and J. Maibaum, *J. Pharm. Sci.*, **74**, 438 (1985).
16. G. Blaschke, H. Kraft, and H. Markgraf, *Chem. Ber.*, **113**, 2318 (1980).
17. G. Blaschke, H. Kraft, K. Fickentscher, and F. Kohler, *Arzneim. Forsch./Drug Res.*, **29**, 1640 (1979).
18. Y. Okamoto, M. Kawashima, R. Aburatani, K. Hatada, N. Nishiyama, and M. Masuda, *Chem. Lett.*, 1237 (1986).
19. Application sheet on Chiracel OD, Daicel, Tokyo.
20. D. Masurel and I. Wainer, *J. Chromatrogr.*, **490**, 133 (1989).
21. I. Fitos, J. Visy, A. Magayar, J. Katjar, and M. Simonyi, *Chirality*, **2**, 161 (1990).
22. F. Reiter, G. Pohl, and H. Grill, Third International Symposium on Chiral Discrimination, Tübingen, Germany, October 1992.
23. P. Erlandson, R. Isaksson, P. Lorenton, and P. Lindberg, *J. Chromatrogr.*, **532**, 305 (1990).
24. L. Miller and H. Bush, *J. Chromatrogr.*, **484**, 337 (1989).
25. P. Salvadori, C. Berucci, E. Domenici, and G. Giannacini, *J. Pharm. Biomed. Anal.*, **7**, 1735 (1989).
26. W. H. Pirkle, A. Tsipouras, and T. J. Sowin, *J. Chromatrogr.*, **319**, 392 (1985).
27. W. H. Pirkle and A. Tsipouras, *J. Chromatrogr.*, **291**, 291 (1984).
28. Chiral AGP, Application sheet, *Chrom Tech*, Stockholm, 1993.

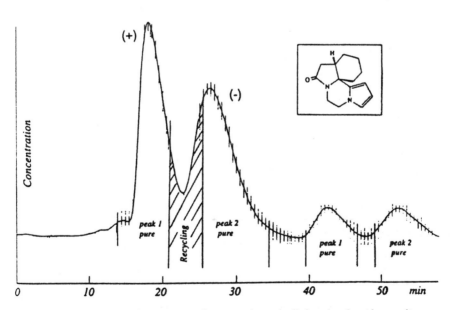

Figure 9.5 Resolution of a nootropic drug on tribenzylcellulose beads with recycling. Reproduced from E. Francotte, in *Chiral Separations: Applications and Technology*, S. Ahuja, Ed., American Chemical Society, Washington, DC, 1997. Copyright 1997 American Chemical Society.

Table 9.4 Preparative Resolution of Racemic Pesticide and Pheromones

Structure	Details	Structure	Details
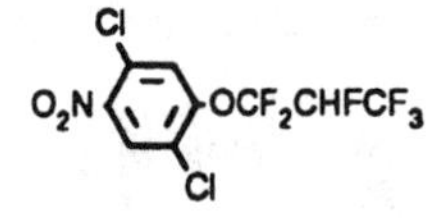 O₂N–C₆H₂Cl₂–OCF₂CHFCF₃	*Activity:* insecticide (intermediate) *Sample size:* 200 mg *CSP:* TBC *Column (cm):* 5 × 75	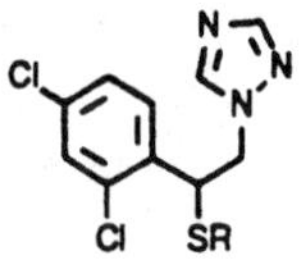 Cl–C₆H₃Cl–CH(SR)–CH₂–triazole	*Activity:* fungicide *Sample size:* 2 mg *CSP:* CTA-1 *Column (cm):* 6.3 × 66
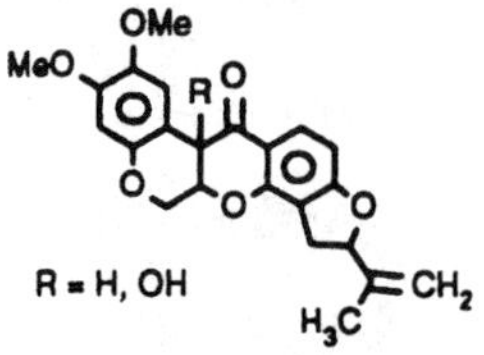 MeO, OMe-substituted chromene, R = H, OH	*Activity:* insecticide *Sample size:* 5 mg *CSP:* PTrMA *Column (cm):* 1 × 25	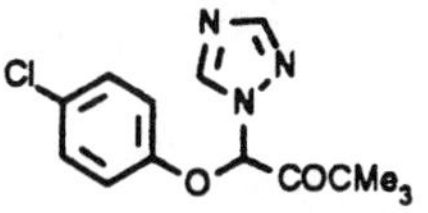 Cl–C₆H₄–O–CH(triazole)–COCMe₃	*Activity:* fungicide *Sample size:* 30 mg *CSP:* CTA *Column (cm):* 2.5 × 30
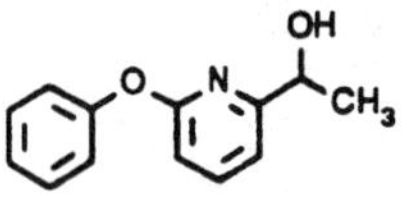 Phenoxy-pyridyl-CH(OH)CH₃	*Activity:* insecticide (intermediate) *Sample size:* 300 mg *CSP:* CTA-1 *Column (cm):* 5 × 60	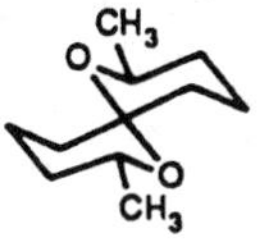 Dimethyl dioxane	*Activity:* pheromone *Sample size:* 140 mg *CSP:* CTA-1 *Column (cm):* 2.5 × 60
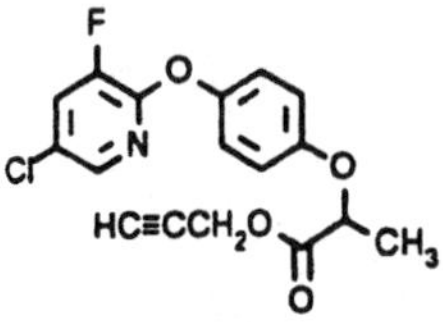 F, Cl pyridyl–O–C₆H₄–O–CH(CH₃)–C(O)–O–CH₂C≡CH	*Activity:* herbicide *Sample size:* 4 mg *CSP:* CTA-1 *Column (cm):* 5 × 60	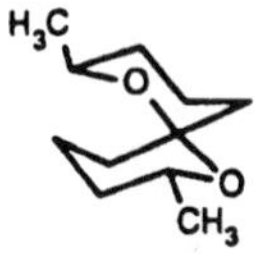 Dimethyl dioxane isomer	*Activity:* pheromone *Sample size:* 80 mg *CSP:* CTA-1 *Column (cm):* 2.5 × 60
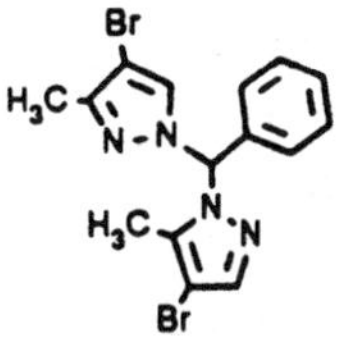 Br, H₃C-pyrazolyl–CH(Ph)–pyrazolyl-CH₃, Br	*Activity:* fungicide *Sample size:* 100 mg *CSP:* CTA-1 *Column (cm):* 2.5 × 20	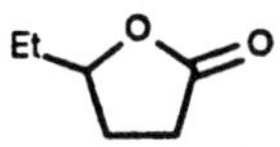 Et-γ-butyrolactone	*Activity:* pheromone *Sample size:* 2.47 g *CSP:* CTA-1 *Column (cm):* 5 × 60
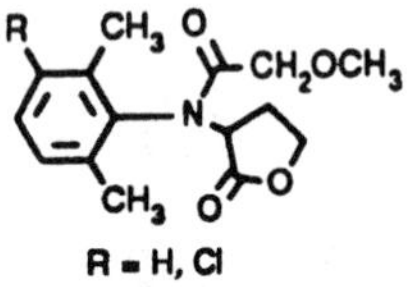 R, CH₃-aryl–N(CH₂OCH₃)–lactone, R = H, Cl	*Activity:* fungicide *Sample size:* 1–52 mg *CSP:* CTA-1 *Column (cm):* 20 × 100		

Source: Reproduced from E. Francotte in *Chiral Separations: Applications and Technology*, S. Ahuja, Ed., American Chemical Society, Washington, DC, 1997. Copyright 1997 American Chemical Society.

Table 9.5 Preparative Separation of the Enantiomers of Compounds Containing Chiral Heteroatoms

Structure	Conditions	Structure	Conditions
(bicyclic diamine with CH₃ groups)	Sample size: 32 mg/65 g CSP: CTA-1 Column (cm): 20 × 100	benzyl phenyl sulfoxide	Sample size: 100 mg CSP: DACH-DNB Column (cm): 5 × 76
H₃C-N(CH₃)(N-CH₃)-CH₂-C₆H₅	Sample size: 100 mg CSP: CTA-1 Column (cm): 2.5 × 41	HN-C₆H₄-OCH₃ with O=S-CH₂COOC(CH₃)₃	Sample size: 200 mg CSP: DNB-Tyr-A Column (cm): 4 × 26
H₃C-C(CH₃)-NH-N-CH₂-C₆H₅	Sample size: 149 mg CSP: CTA-1 Column (cm): 2.5 × 41	thiadiazine with C₆H₅, C₆H₅, CH₃, N=O	Sample size: 15 mg CSP: Poly-PEA Column (cm): 1 × 24
H-N=N-C(CH₃)₂-CH₂-CH₂-C₆H₅	Sample size: 100 mg CSP: CTA-1 Column (cm): 2.5 × 30	naphthyl-P(=O)(CH₃)(OCH₃)	Sample size: 500 mg/1 g CSP: DNBPG-co Column (cm): 8 × 22
benzyl-N-NH-cyclohexyl	Sample size: 120 mg CSP: CTA-1 Column (cm): 2.5 × 30	bicyclic phosphorus (CH₃, Ph, Ph, COOEt)	Sample size: 100–150 mg CSP: DNB-Tyr-E Column (cm): 4 × 26
(C₆H₅)₂C-N(Cl)-N-H	Sample size: 100 mg CSP: MMBC Column (cm): 2.5 × 50	bicyclic phosphorus (CH₃, Ph, Ph, CH₂OH)	Sample size: 100 mg CSP: DNB-Tyr-A Column (cm): 6 × 7.6
Et-N(Me)-naphthyl-C(OH)(C₆H₅)(C₆H₅)	Sample size: 40 mg CSP: CTA-1 Column (cm): 1.2 × 50	cyclopentanone-P(=O)(C₆H₅)(C₆H₅)(C₆H₅)	Sample size: 38 mg CSP: DNB-Tyr-E Column (cm): 6 × 10
H₂N-C(=O)-CH₂-CH₂-S(=O)₂-C₆H₅	Sample size: 100 mg CSP: Chiralcel OB Column (cm): 2 × 50	tBu-phenyl selenoxide	Sample size: 220 mg CSP: DNBPG-co Column (cm): 1.1 × 30
anthracene-S(=O)-C₄H₉, CH₂Cl	Sample size: 3 g CSP: DNBPG-io Column (cm): 5 × 76	(C₆H₅)₃C-Sn(CH₃)(C₆H₅)-C(CH₃)(C₆H₅)(CH₃)	Sample size: 100 mg CSP: CTA-1 Column (cm): 2.8 × 6.6

Source: Reproduced from E. Francotte in *Chiral Separations: Applications and Technology*, S. Ahuja, Ed., American Chemical Society, Washington, DC, 1997. Copyright 1997 American Chemical Society.

Table 9.6 Preparative Separation of the Enantiomers of Planar Chiral Compounds

Structure	Conditions	Structure	Conditions
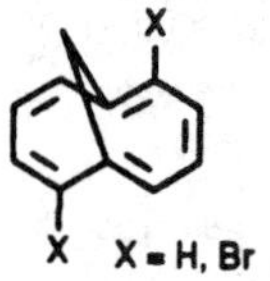 X = H, Br	Sample size: 10–450 mg CSP: CTA-I Column (cm): 6.3 × 66	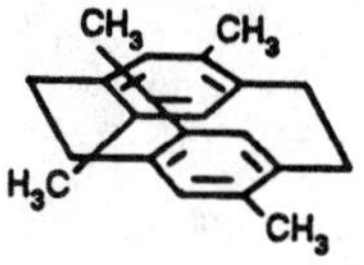 (hexamethyl [2.2]paracyclophane)	Sample size: 70 mg CSP: PTrMA Column (cm): 0.95 × 37.5
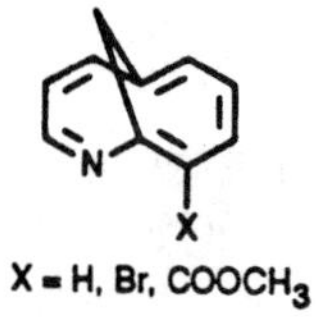 X = H, Br, COOCH$_3$	Sample size: 10–20 mg CSP: CTA-I Column (cm): 2.5 × 30	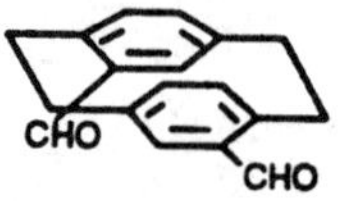 diformyl [2.2]paracyclophane	Sample size: 5 mg CSP: Chiralcel OD Column (cm): 0.46 × 25
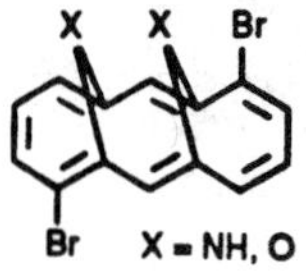 X = NH, O; Br substituents	Sample size: 20 mg CSP: CTA-I Column (cm): 2.5 × 30	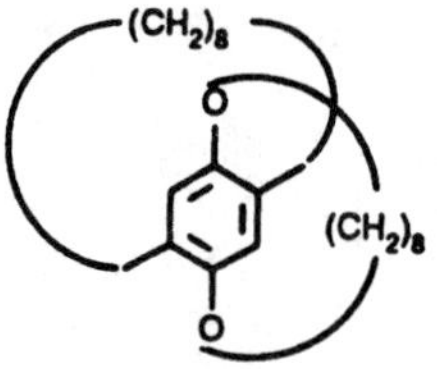 (CH$_2$)$_8$ bridged cyclophane	Sample size: 70 mg CSP: PTrMA Column (cm): 0.95 × 37.5
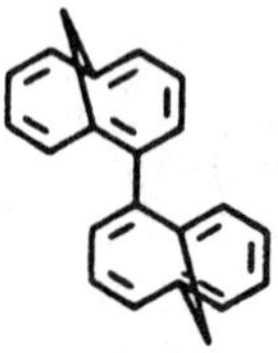 binaphthyl	Sample size: nm. CSP: CTA-I Column (cm): 6.3 × 66	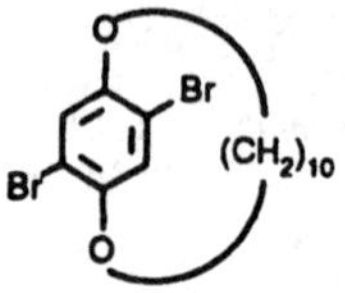 Br, Br with (CH$_2$)$_{10}$	Sample size: 250 mg CSP: CTA-I Column (cm): 3.1 × 40
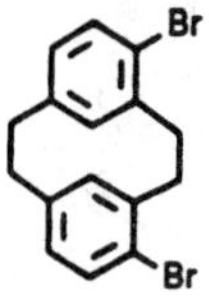 dibromo [2.2]paracyclophane	Sample size: 5–500 mg CSP: CTA-I Column (cm): 6.3 × 66	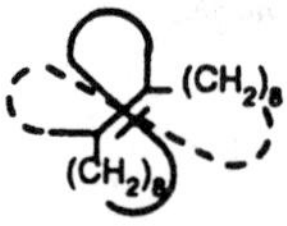 (CH$_2$)$_8$, (CH$_2$)$_8$	Sample size: 200 mg CSP: PTrMA Column (cm): 1.16 × 80.6
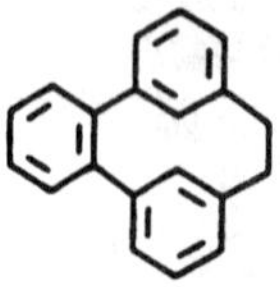 [2.2]metaparacyclophane	Sample size: 50 mg CSP: CTA-I Column (cm): 2.5 × 30		

Source: Reproduced from E. Francotte in *Chiral Separations: Applications and Technology*, S. Ahuja, Ed., American Chemical Society, Washington, DC, 1997. Copyright 1997 American Chemical Society.

Other methods

GC methods. Although GC has been used for a number of applications on the analytical scale, there are a limited number of applications in the preparative scale. A few examples are given in Table 9.8.

Most of the separations have been achieved on cyclodextrin-based CSPs obtained by coating a dilute solution of cyclodextrin or its derivative on Chromosorb. The examples include pheromones, monoterpenes, and anesthetics.

Table 9.7 Preparative Resolution of Racemates with Helical or Propeller Chirality

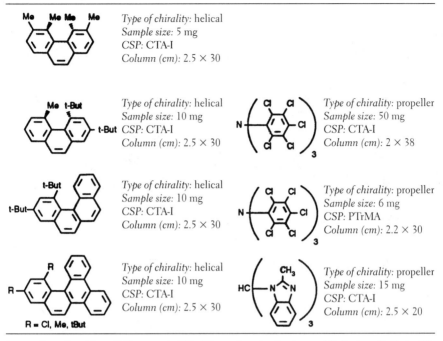

Source: Reproduced from E. Francotte in *Chiral Separations: Applications and Technology*, S. Ahuja, Ed., American Chemical Society, Washington, DC, 1997. Copyright 1997 American Chemical Society.

SFC methods. Pilot-scale separations have been demonstrated with SFC (see Table 9.9). Because of low viscosity of the mobile phase (mainly carbon dioxide), SFC can be performed at high flow rates and is generally faster than HPLC.[12] This helps provide greater production. A number of examples have been summarized in Table 9.9.

It should be noted that the limited solubility of most drugs and agrochemicals in carbon dioxide considerably reduces the applicability of this technique. Solubility can be improved by the addition of a limited amount of modifiers such as alcohols; however, their elimination presents additional problems.

Large-scale methods. The feasibility of separating enantiomers up to the kilogram scale by chromatography on CSPs has been established. Techniques such as peak shaving and recycling can help improve yields and production. The development of techniques such as simulated moving-bed (SMB) chromatography and membrane separations will contribute greatly in this area. A few examples of separations by SMB are given in Table 9.10. An additional advantage of SMB is that it can help save up to 90% of the mobile phase.

Table 9.8 Preparative Chromatographic GC Resolutions

Racemate	Chiral Selector	Column (m × cm)	Sample Size (mg)	Carrier Gas	Flow rate
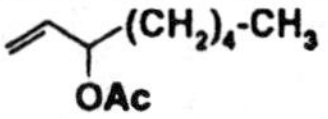 CH₂=CH-CH(OAc)-(CH₂)₄-CH₃	β-Cyclodextrin	1 × 2	430	Hydrogen	2.6 L/min
$CF_3CHClOCHF_2$	Trifluoroacetyl-γ-cyclodextrin	2 × 1	75-675	Helium	2.2 cm/s
$CF_3CHClOCHF_2$	2,6-Di-O-pentyl-3-O-butanoyl-γ-cyclodextrin	4 × 0.7	6	Helium	8.4 cm/s
$CHFClCF_2OCHF_2$	Trifluoroacetyl-γ-cyclodextrin	1 × 1	50-120	Hydrogen	5 cm/s
$CHFClCF_2OCHF_2$	2,6-Di-O-pentyl-3-O-butanoyl-γ-cyclodextrin	4 × 0.7	47	Helium	8.4 cm/s
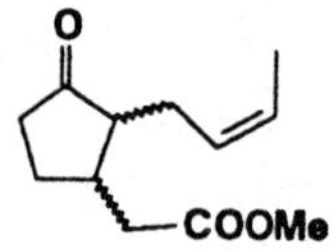 (cyclopentanone with CH₂-COOMe and allyl substituents)	2,6-Di-O-methyl-3-O-pentyl-β-cyclodextrin	1.8 × 0.4	2	Helium	400 mL/min
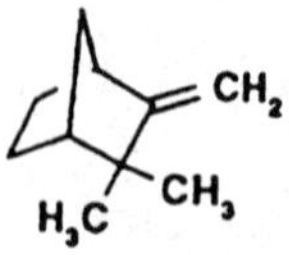 (norbornane with =CH₂ and gem-dimethyl)	α-Cyclodextrin	2 × 0.4	—	Helium	—
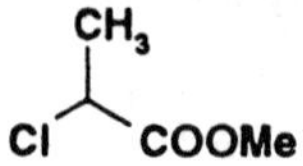 CH₃-CCl(H)-COOMe	Trichloroacetyl-β-cyclodextrin	1 × 2.25	206	Hydrogen	8 cm/s
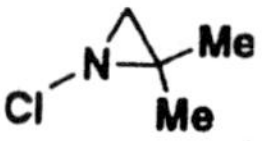 Cl-N(aziridine)-C(Me)₂	Nickel(II) bis [(3-heptafluorobutanoyl)-(1R)-camphorate]	7.5 × 0.3	2	Nitrogen	1.5 bar

Source: Reproduced from E. Francotte in *Chiral Separations: Applications and Technology*, S. Ahuja, Ed., American Chemical Society, Washington, DC, 1997. Copyright 1997 American Chemical Society.

Table 9.9 Preparative Chromatographic Resolutions with SFC

Racemate	CSP	Column Dimensions (cm)	Sample Size (mg)	Mobile Phase	Flow Rate (mL/min)
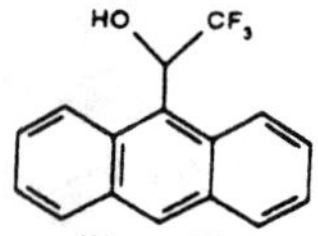 9-(anthryl)-CH(OH)(CF₃)	DNBPG-co	25 × 0.9	10	CO_2-propanol (82:18)	8
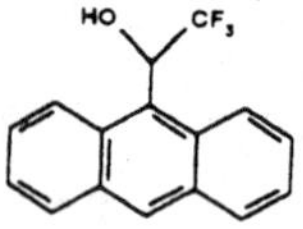 9-(anthryl)-CH(OH)(CF₃)	ChyRoSine-A	7.6 × 6	40	CO_2-ethanol (96.5:3.5)	717

(continued)

Table 9.9 (Continued)

Racemate	CSP	Column Dimensions (cm)	Sample Size (mg)	Mobile Phase	Flow Rate (mL/min)
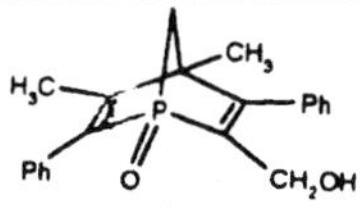 (structure)	DNB-Tyr-E	7.6 × 6	100	CO_2-ethanol (92:8)	167
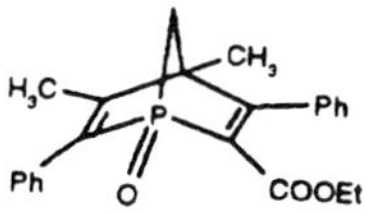 (structure)	DNB-Tyr-E	10 × 6	100	CO_2-ethanol (88:12)	200
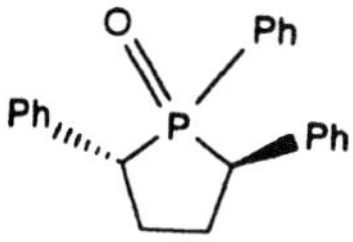 (structure)	DNB-Tyr-E	10 × 6	38	CO_2-ethanol (92:8)	250
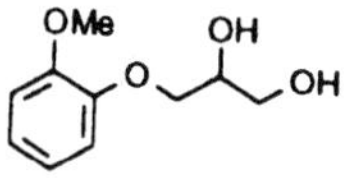 (structure)	Chiralcel OD	23 × 1	60	CO_2-methanol-2-propanol (86:7:7)	5
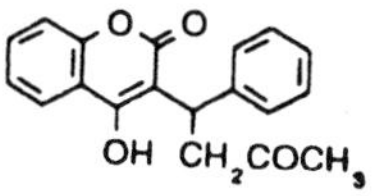 (structure)	Whelk O-1	2 × 2.54		CO_2-2-propanol-AcOH (75:25:0.5)	100

Source: Reproduced from E. Francotte in *Chiral Separations: Applications and Technology*, S. Ahuja, Ed., American Chemical Society, Washington, DC, 1997. Copyright 1997 American Chemical Society.

Table 9.10 Preparative SMB

Compound	CSP	Productivity
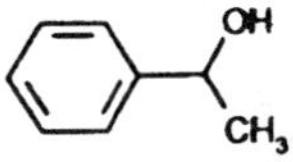 (structure)	Chiracel OD	0.98 g/hr
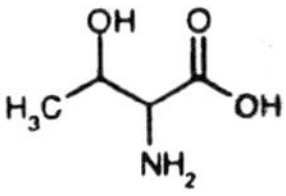 (structure)	Chirosolve L-proline	Not available
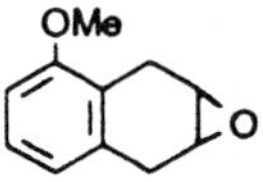 (structure)	CTA-I	1.45 g/hr
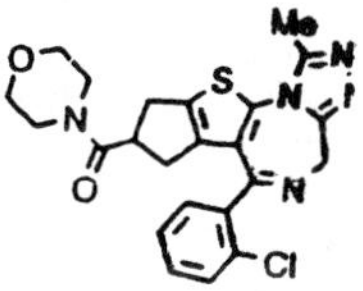 (structure)	CTA-I	119 g/hr

(continued)

Table 9.10 (Continued)

Compound	CSP	Productivity
(tetrahydroquinoline-thiazolidinone structure)	PMBC	4.1g/hr
(naphthyloxy-propanol-amine structure)	Chiralcel OD	3.8 g/hr

Source: Reproduced from E. Francotte in *Chiral Separations: Applications and Technology*, S. Ahuja, Ed., American Chemical Society, Washington, DC, 1997. Copyright 1997 American Chemical Society.

The usefulness of membrane separations has been discussed.[1] Details of separation conditions have been provided for phenylpropanolamine, ephedrine, metazapine, phenyl glycine, albuterol, terbutaline, propanol, and ibuprofen.[13]

Figure 9.6 demonstrates the use of a hollow-fiber membrane reactor for separation of an L-acid from a racemic ester. Hollow-fiber membrane laboratory-scale modules have demonstrated the capacity of resolving 165 kg per year for a compound of molecular weight of 150. With 20 full-scale modules, a capacity in excess of 100,000 kg per year can be obtained.

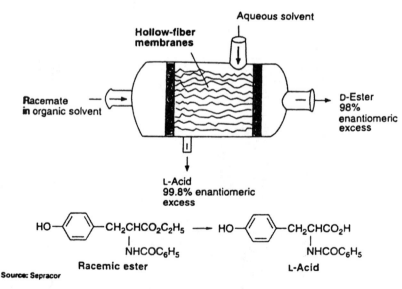

Figure 9.6 Preparation of an L-acid.

REFERENCES

1. S. Ahuja, *Chiral Separations: Applications and Technology.* American Chemical Society: Washington, DC, 1997.
2. E. Francotte, in *Chiral Separations: Applications and Technology,* S. Ahuja, Ed. American Chemical Society: Washington, DC, 1997.
3. S. Ahuja, W. Pirkle, and C. Welch, *Chiral Separations by Chromatography,* American Chemical Society Short Course, 1994.
4. R. M. Nicoud, G. Fuchs, P. Adams, M. Bailey, E. Kusters, F. D. Antia, R. Reuille, and E. Schmid, *Chirality,* **5,** 267 (1993).
5. C. Berger and M. Perrut, *J. Chromatogr.,* **505,** 37 (1990).
6. C. Roussel, J. Stein, F. Beauvais, and A. Chemlal, *J. Chromatogr.,* **462,** 95 (1989).
7. N. Krause and G. Handke, *Tetrahedron Lett.,* **32,** 7225 (1991).
8. M. Okamoto and H. Nakazawa, *J. Chromatogr.,* **588,** 177 (1991).
9. K. Balmer, P. Lagestrom, and B. Persson, *J. Chromatogr.,* **592,** 1331 (1992).
10. E. Francotte and P. Reicher, unpublished, through ref. 2 of this chapter.
11. K. Schlogl and M. Widhalm, *Monash. Chem.,* **115,** 1113 (1984).
12. E. Francotte. In *Chiral Separations: Applications and Technology,* S. Ahuja, Ed. American Chemical Society: Washington DC, 1997, p. 271.
13. L. J. Brice and W. H. Pirkle, In *Chiral Separations: Applications and Technology,* S. Ahuja, Ed. American Chemical Society: Washington, DC, 1997, p. 309.

REVIEW

1. Define displacement and elution chromatography.
2. What are the advantages of using displacement chromatography?
3. List modes of chromatography that are commonly used for preparative separations.
4. List at least two CSPs that have been found most useful for preparative separations.

TUTORIAL

The choice of preparative chromatography is likely to be dictated largely by the type of sample and the amount of desired enantiomer to be isolated. For very small amounts, even capillary electrophoresis can be used (e.g., see Micro-preparative Applications of Capillary Electrophoresis, *Isolation and Purification,* **2,** 113 [1996]). For volatile samples, GC is the technique of choice. HPLC is likely to be favored for most other samples. A number of columns have been used for these separations; cellulose columns are generally favored. Of these, cellulose triacetate (CTA-I) is the most popular. π-Acid π-base columns have been found useful because they offer high α values and allow reversing the peak order. SFC provides advantages of both GC and HPLC; however, the primary limitation is solubility of a given sample in carbon dioxide (the commonly used mobile phase). The solubility can be improved by addition of modifiers, such as alcohol in limited amounts; however, this presents elimination problems during collection of separated materials. For large-scale separations, SMB has been found very useful; however, a significant amount of capital and know-how are necessary for obtaining successful separations. Membrane separations offer a significant advantage, but again, expertise and experience are essential.

10

Method Selection and Selected Applications

This chapter is designed to enable readers to select the best method for their applications. It starts by showing how to determine whether the molecule of interest has the required asymmetry to confer chirality and works through selection of a suitable chromatographic method. The material then helps with the selection of a suitable column for HPLC methods (the methods most commonly used for chiral separation), details applications based on structure, and, finally, provides selected applications with other methods, such as GC, SFC, and CE.

Where to Start?

The hardest problem for an investigator in chiral separations is to determine where to start. The experts in this area may want to skip to the next section. Experts seem to talk in a language of their own; they claim they can easily carry out separations of chiral compounds, but they don't elaborate on the basic rationale for achieving these separations. Frequently, they tend to favor a certain type of column — for example, brush-type columns or cyclodextrin columns. They don't mention that in each case a different mode of separation may be involved and that different handling steps are necessary. Furthermore, many experts tend to lead everyone to the area with which the experts are most familiar. This, of course, is of no help to the beginner because he or she wants to know a good place to start and how to make rational decisions as to the best mode of separation.

Cost considerations, availability of equipment, and know-how play important roles in the selection process.[1, 2] Based on earlier discussions in this book, it should be apparent that chromatographic methods are likely to be favored over nonchromatographic methods. The next step is choosing between TLC, GC, HPLC, SFC, and CE (see Which Method to Use? later in this chapter). For most separations, HPLC methods are likely to be useful (exceptions have been noted in chapter 4). The next step is to select an appropriate column based on considerations given in chapters 7 and 8. The primary considerations for any operator are as follows, notwithstanding high cost of the columns:

- Will the column work for my application?
- Is the column relatively easy to work with?
- Will it be ruined easily?

This chapter attempts to answer these questions and provides examples that will help the reader make intelligent decisions in this complex field.

Study the Molecule

First the investigator must take a close look at the molecule that is to be separated and then answer the following question: Is there a stereogenic center? The simplest example of this is an asymmetric carbon with four different substituents (Figure 10.1).

As simple as it sounds, this process often appears cumbersome when one is looking at many carbon atoms in a molecular structure. The simplest way to counter this problem is to number all carbon atoms in the structure and look at each of them in turn to see if they are asymmetric (see Figure 10.2). If no asymmetric carbon is found, one should look at the plane of symmetry of the whole molecule and/or other atoms, such as sulfur and nitrogen that can also confer chirality (see chapter 3). If one asymmetric carbon or stereogenic center is found, the investigator can expect two enantiomers. For two stereogenic centers, the number of enantiomers is four. It should be clear that this number grows rapidly as the number of asymmetric centers increases; the 2^n Rule applies, except for fused rings, where n is the number of asymmetric centers.

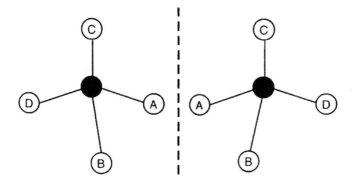

Figure 10.1 An asymmetric C compound.

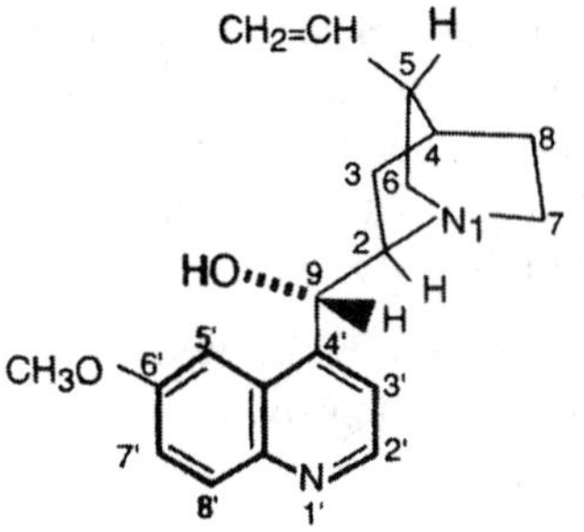

Figure 10.2 Numbering C atoms to determine asymmetry.

Which Method to Use?

Once it is clear that chirality exists and enantiomers exist, the next rational step is to look at the structure of the whole molecule to determine what type of interactions are likely to come into play to yield a useful separation. At this stage, it is also necessary to decide which method(s) should be used. The options in chromatographic methods are TLC, GC, HPLC, CE, and SFC. In my opinion, if qualitative or semiquantitative information is desired and cost of equipment is a factor, TLC should be tried (chapter 4). Of course, TLC can also assist with the method development for HPLC because of similarities in the mode of separation. Gas chromatography is very useful for volatile compounds (see chapters 4 and 9). If derivatization is necessary to make the solute volatile, HPLC becomes the technique of choice unless the equipment or other considerations bias the experimenter toward GC.

HPLC is a popular technique that offers ease of operation; derivatization is frequently not required, and the number of possible separation modes is large. Therefore, the discussion below focuses on HPLC methods. Other methods such as CE and SFC can be used, depending on the interest of the reader and availability of instrumentation. Their advantages and limitations have been discussed earlier in this book. Some additional information is provided later in this chapter.

High Pressure Liquid Chromatography

The flowchart shown in Figure 10.3[3] depicts pathways to separation of enantiomers in HPLC, and descriptions of various types of CSPs can be found in Table 8.2. The choice of CSP can be based on whether normal phase or reversed phase is favored for a given separation (see Figure 8.1). Further discussion on this subject is given in the next subsection.

Choice of Stationary Phase

As mentioned, it is important to study the solute to determine what kind of interactions it can engender. This information may help in the selection of the de-

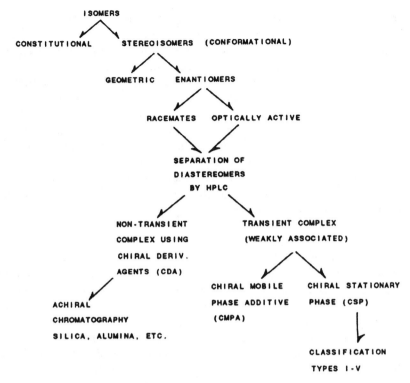

Figure 10.3 Flowchart for separating isomers. Reproduced with permission from Innovative Directions in Chiral Separations, *Phenomenex Bulletin*, Torrance, CA. Copyright Phenomenex.

sired stationary phase. For a quick review of various interactions on CSPs, see Table 8.1.

Brush-type columns (Type 1). It is important to determine if the molecule of interest is likely to form hydrogen bonds, have π-π interactions, have dipole stacking, or have other attractive interactions.

The chiral stationary phase (CSP) in brush-type columns is composed of various selectors capable of ionic or covalent bonding. Enantiomeric compounds that can be separated on these columns are as follows:

Amines	Ethers
Amino acids	Hydroxy acids
Carboxylic acids	Ketones
Esters	Lactones

We know that brush-type CSPs are generally composed of an optically pure amino acid bonded to γ-aminopropyl-silanized silica gel. An amide or urea linking of a π electron group to the asymmetric center of amino acid provides for π-electron interactions and one point of chiral recognition. It has been proposed that at least three points of interaction are necessary between the chiral molecule and the CSP. One of these interactions must be stereochemically dependent. When three points of interaction occur, a transient diastereomeric complex is formed between the solute and the CSP. The π acids, π bases, hydrogen bond donors and/or acceptors, amide dipoles, and other functional groups linked to a chiral selector of the CSP provide the required three-point interaction and confer overall selectivity to the column. Other points that provide attractive or repulsive interactions also contribute to enantioselectivity. If the sample of interest does not contain the necessary sites of recognition, they have to be added by forming appropriate derivatives—a process not very popular with separation scientists. Two types of columns commonly used are discussed below.

Amide-type columns. In amide-type columns, a chiral selector such as phenylglycine or 1-naphthylglycine is derivatized to form an amide linkage with the π acid (electron acceptor) dinitrobenzoic acid (DNB) (Figure 10.4). DNB is responsible for π-π interactions with electron-rich π basic groups on the analyte molecule and act as one point in the chiral recognition process. Chiral analytes containing phenyl, naphthyl, and anthryl groups are strong π bases and show increasing chiral recognition and selectivity as the strength of the π acid increases. Other points of interaction are secondary hydrogen bonding with the DNB carbonyl and binding interactions with the amide functionality. The amide group has the ability to serve as either a donor or acceptor for the purpose of hydrogen bonding.

Urea-type columns. Chiral selectors such as valine, proline, leucine, or indoline-2-carboxylic acid are derivatized through the urea linkage with either the π acid dinitroaniline or the π base 1-(α-naphthyl) ethylamine to make up the CSP. The mechanism of chiral recognition is similar to amide chiral phases discussed earlier. Aromatic groups on the CSP interact with the aromatic groups of the enantiomers, providing one point of interaction by either donating or withdrawing electrons. Additional hydrophobic interactions are derived from the ethyl group of 1-(α-naphthyl) ethylamine or other allyl substituents of the CSP. The urea group can form hydrogen bonds or allow other dipole-dipole interactions. The interactions of the groups in the CSP with those of the chiral analyte lead to short-lived diastereomeric pairs with different energies and stability. Good resolution is obtained when the difference in energy (stability) is large.

It has been theorized that separation of the drug propanolol on a urea-type column based on valine is dependent on the stereoselective interactions of the chiral hydroxyl proton, with the carbonyl function in the urea linkage of the CSP. Significant intermolecular interactions also occur between the π electrons of the two naphthyl groups. Acidic N-H protons are present on both the CSP and analyte hydrogen bond with the opposing basic oxygen atoms (the ether and carbonyl functions). The steric interaction of the valine moiety also adds to the overall enantio-

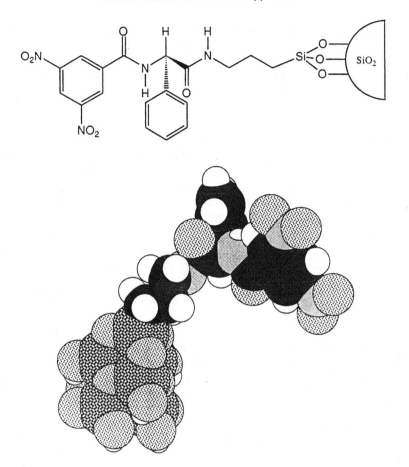

Figure 10.4 DNB-phenylglycine CSP.

selectivity of the separation. Conformational and steric differences, combined with hydrogen bonding and electrostatic interactions, lead to diastereomers of different stability, thus making the separation possible.

To summarize, with brush-type CSPs, the chiral recognition process is believed to be dependent on three points of interaction between the analyte and the CSP. The interactions of each enantiomer with the CSP are rapid and reversible, so that separation is directly related to the energy difference between the transient diastereomeric pair. The greater the energy difference, the greater the resolution. Highly specific interactions in the chiral environment result in a high degree of selectivity. The enantiomer with the most stable association with the CSP will be retained longest on the column. This means it is possible to reverse the elution order of the racemate by simply employing the opposite chiral phase.

Operation mode. Amide and urea columns are covalently bonded and are highly stable under both normal- and reversed-phase conditions.

1. Normal-Phase Separations. Normal-phase separations generally provide better enantioselectivity. Hexane and 1,2-dichloroethane are typically used as solvents for normal-phase separations. For chiral separations requiring polar modifiers, hexane and dichloroethane can be modified with methanol, ethanol, isopropanol, tetrahydrofuran, chloroform, ethyl acetate, or trifluoroacetic acid. A change in type and concentration of alcohols can effectively control retention times and resolution. Acetonitrile (up to 2%) can be used in the mobile phase to decrease retention time and improve efficiency; however, some loss in resolution is likely. A small amount of competing amines (0.1% to 0.5%) or acid added to the mobile phase can improve resolution of acidic and basic enantiomers.

2. Reversed-Phase Separations. Reversed-phase separation can be useful for ionic and highly polar analytes such as carboxylic acids. The mobile phase is composed of salts such as ammonium acetate dissolved in methanol-water. Acetonitrile and tetrahydrofuran can be used as additional modifiers. These columns should generally be operated at temperatures below 50°C; resolution of enantiomeric pairs can in some cases be enhanced by using subambient temperatures. Low flow rates up to 1 mL/minute are favored.

3. Achiral Derivatization. Chiral recognition is optimum when the three chiral recognition sites are adjacent to the stereogenic center.[4] If a molecule contains the necessary recognition characteristics for the common brush-type column, derivatization is usually not required. Examples include the β-binaphthols, (s)-aryl-β-hydroxyl sulfoxides, C-aryl hydantoins, C-aryl succinimides, and indoline 2-carboxylates.

Derivatization with an achiral agent can provide additional interaction sites, and achiral derivatization can be selected to enhance detectability as well. For example, derivatization is typically required for analytes lacking aromatic substituents. While many analytes contain π-basic aryl groups that can be resolved on π-acid CSPs without derivatization, relatively few analytes possess π-acidic groups. Incorporation of such a group by derivatization is usually the first step if a π-donor column is to be used. This entails reaction with reagents such as 3,5-dinitrobenzoyl chloride, 3,5-dinitrobenzylisocyanate, or 3,5-dinitroaniline. In the case of both type of columns, derivatization can reduce analysis time and improve band shape. For example, amines are often acylated, carboxylic acids are esterified or converted to amides or anilides, and alcohols are sometimes converted to carbamates or esters.

Derivatization of groups remote from the stereogenic center can improve resolution by reducing the extent to which these groups contribute to achiral retention. Table 10.1 provides a selection of appropriate achiral derivatization reagents for various compounds. For details on derivatization reaction conditions, the reader may want to review reference 4.

4. Underivatized Separations. Several new columns such as Whelk-O 1 or α-Burke columns have evolved out of research on the brush-type of columns, which obviate the need for derivatization. Whelk-O 1 is a useful CSP for HPLC separa-

Table 10.1 Derivatization for Brush-Type Phases

Analyte	Reagent	Derivative
Alcohols		
Aliphatic/Aromatic	N-imidazole-N'-carbonic acid-3,5-dinitroanilide or 1-naphthylisocyanate	Carbamate
Amines		
Aliphatic/Aromatic	N-imidazole-N'-carbonic acid-3,5-dinitroanilide	Urea
	2-naphthoyl chloride	Amide
Amino Acids		
Aromatic/Aliphatic	3,5-dinitrobenzoyl chloride/any alcohol	Amide/ester
	Acetyl chloride/any alcohol	Amide/ester
Carboxylic Acids		
Aromatic/Aliphatic	3,5-dinitroaniline	Amide
Peptides	3,5-dinitrobenzoyl chloride	Amide
Piperidine, 2-methyl	1-naphthylisocyanate	Amide
Thiols	N-imidazole-N'-carbonic acid-3,5-dinitroanilide	Carbamate

Source: Reproduced from S. Perrin and W. Pirkle, in *Chiral Separations by HPLC*, S. Ahuja, Ed., American Chemical Society Symposium Series #471, Washington, DC, 1991. Copyright 1991 American Chemical Society.

tion of enantiomers of a wide variety of compounds. This column was originally developed for separating the enantiomers of naproxen. The presumed interactions of the CSP with the two enantiomers of naproxen are shown in Figure 10.5. A similar type of interaction with the CSP is believed to account for the ability of this column to resolve the enantiomers of analyte having a π-aromatic system with a hydrogen bond acceptor located near the stereogenic center (Figure 10.6)

With the variety of separations studied with this column, the groups participating in π-π and hydrogen bonding interactions have been identified. This column also offers high stability, good preparative capacity, and the ability to be used in the normal- and reversed-phase modes.

Naproxen can be resolved using a normal-phase method with hexane/IPA/HOAc (80/20/0.5) on Whelk-O 1 (4.6 mm i.d. × 25 cm) at a flow rate of 1 mL/min in 16 minutes.[5] Alternatively, methanol:0.1% phosphate (60:40) can be used at the same flow rate with the same run time. The α value is 2.1 in normal-phase and 1.7 in reversed-phase mode. Other arylpropionic acids that have been resolved in the normal-phase mode are ibuprofen, ketoprofen, and flubiprofen. A higher concentration of hexane and a correspondingly lower concentration of IPA were employed for these separations. Other pharmaceuticals that have been resolved on this column are cyclothiazide, bendroflumethiazide, oxazepam, mephenytoin, bupivacaine, and p-chloro-warfarin. The α-Burke II column is useful for resolution of metoprolol and related compounds.

A mobile phase composed of methylene chloride:ethanol:methanol (85:10:5) containing 10 mm ammonium acetate is used at a flow rate of 1 mL/minute for a 4.6 mm i.d. × 25 cm column. The α value of the enantiomers is 1.28. Related com-

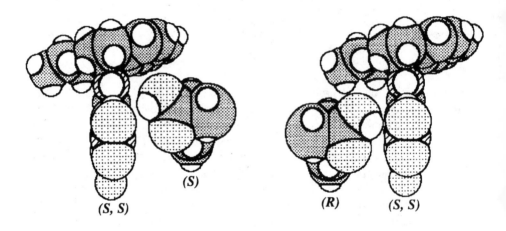

Figure 10.5 Whelk-O 1 CSP.

pounds such as alprenolol, atenolol, betaxolol, bufralol, bupranolol, oxprenolol, practolol, pindolol, pronethalol, and propanolol can be resolved with slight modifications of mobile phases.

Inclusion CSPs (Type 3)

Cyclodextrin columns. Recall that cyclodextrins are produced by the partial degradation of starch and the enzymatic coupling of cleaved glucose units into crystalline, homogeneous toroidal structures of different molecular size. Alpha-, β-, and γ-cyclodextrins have been most widely characterized; they contain six, seven, and eight glucose units, respectively, and are chiral (Figure 10.7). For example, β-cy-

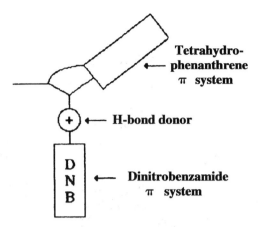

Figure 10.6 Schematic diagram showing key functional groups of Whelk-O 1.

clodextrin has 35 stereogenic centers, and the toroidal structure has a hydrophilic surface resulting from the 2-, 3-, and 6-position hydroxyl groups making them water-soluble.

The cavity is composed of the glucoside oxygen and methylene hydrogen, giving it an apolar character. As a result, cyclodextrins can include polar molecules of appropriate dimensions in their cavities and bind them through dipole-dipole interactions, hydrogen bonding, or London dispersion forces.

X-ray crystallographic data show that the structures of β- and γ-cyclodextrins are quite rigid and inflexible. The bonds of α-cyclodextrin appear to be weak and may possibly be stretched. This means that a larger molecule could fit in if the cavity of α-cyclodextrin is deformed under high aqueous buffer conditions. However, this is not the case with β- and γ-cyclodextrins. In general, cyclodextrins are stable from pH 3 to 14. The physical properties of cyclodextrins are given in Table 10.2.

The water solubility of β-cyclodextrin increases more than 20-fold if hydroxy groups in position 2 or 3 of β-cyclodextrin are ethoxylated or propoxylated.

One or two of the primary hydroxyl groups are used to link the cyclodextrin to the surface of media. The secondary hydroxyl groups can be derivatized selectively, generally in positions 2 and 3. As substitution increases, the reaction rate decreases. Several derivatives of native cyclodextrins have been made to change physical and chemical properties of the cyclodextrins.

Beta-cyclodextrin has found the largest number of applications because it has been found useful for low molecular weight analytes in the pharmaceutical and environmental areas. A number of derivatives have been prepared for HPLC to accommodate various special requirements and to provide specific interactions with certain functional groups in order to produce highly selective separations for a versatile array of analytes. Some of the derivatized cyclodextrins that have proved to be useful are: R,S,-hydroxypropyl ether, S-naphthylethyl carbamate, and R-naphthylethyl carbamate.

α-cyclodextrin

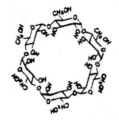

β-cyclodextrin

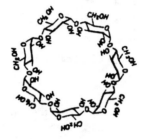

γ-cyclodextrin

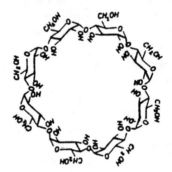

Figure 10.7 Cyclodextrins.

Improved columns such as the second generation of cyclodextrin columns (Cyclobond I 2000, Astec) offer the following advantages: greater stability in aqueous mobile phases, better batch-to-batch reproducibility, and improved selectivity and resolution.

Operation mode. Cyclodextrin columns can be operated in reversed-phase or normal-phase modes.

1. Reversed-Phase Separations. Cyclodextrin columns can be operated in the reversed-phase mode. When cyclodextrin stationary phases are used with aqueous mobile phases, the basic mechanism that controls retention of solutes is called in-

Table 10.2 Physical Properties of Cyclodextrins

Cyclodextrin	Glucose Units	MW	Cavity Diameter (nm)	Water Solubility (g/100 mL)
α	6	972	0.57	15.5
β	7	1135	0.78	1.85
γ	8	1297	0.95	23.2

clusion complexation. This mechanism represents the attraction of the apolar portion of the molecule or the molecule as a whole to the apolar cavity. When an aromatic group is present, the orientation in the cavity is selective because of electron sharing of the aromatic methylene groups with those of the glucoside oxygens. Linear or acyclic hydrocarbons occupy more random positions in the cavity. To achieve a chiral separation in the reversed-phase mode, it is essential that the analyte have at least one aromatic ring. Exceptions are heterocyclic analytes and t-boc amino acids.

The high density of secondary hydroxyl groups at the larger opening of the toroid acts as an energy barrier for polar molecules to complex, and preferential hydrogen bonding occurs. Amines and carboxyl groups interact strongly with these hydroxyl groups as a function of the pKa of the analyte and the pH of the aqueous medium. It is important to understand this relationship if we want to design useful mobile phases for separations of chiral compounds on cyclodextrin columns.

For cyclodextrin inclusion, the molecular weight of a polyaromatic ring structure is not as critical as its bulk. The most important consideration for proper retention and chiral recognition is proper fit of the molecule in the cyclodextrin cavity. This fit is a function of both size and shape of the analyte relative to the cavity. For example, an analyte like norgestrel, a five-ring steroid structure, is better separated on a γ-cyclodextrin column, while the enantiomer of a naphthalene-like structure or single substituted aromatic ring would better fit on a β-cyclodextrin column. Listed here is a general plan that can be used to make initial choices:

- α-Cyclodextrins are useful for small molecules.
- β-Cyclodextrins can be used for substituted phenyl, naphthyl, and binaphthyl ring compounds.
- γ-Cyclodextrins are likely to be useful for molecules with three to five rings in their structure—for example, steroids.

A. *Inclusion Effects.* Inclusion experiments with pyrene have been used to demonstrate both the apolar nature of the cyclodextrin cavity and the relationship to the fit of the analyte. The output of fluorescence indicates the degree of fit of pyrene; greater fit indicates more complete inclusion and a more nonpolar environment. The general polarity of the cavity is comparable to that of the oxygenated solvents such as n-octanoic, n-octanol, isopropyl ether, and t-amyl alcohol.[6] This correlation is applicable to free or bonded cyclodextrins and is independent of cavity size.

The response to mobile phase polarity for the bonded cyclodextrins is roughly equivalent to other C-8 reversed phases. The correlation is not exact in that small water-soluble aromatic molecules show a greater affinity to the cyclodextrin structure than to C-8. This is due to the low capacity of standard reversed-phase separation in high aqueous mobile-phase compositions resulting from the interaction of hydrocarbon moieties to the exclusion of other analytes. However, for the purpose of selecting a mobile phase, the correlation is adequate.

B. *Solvent Strength.* Water is the weakest solvent, in terms of its interactions with the cyclodextrin cavity. Alcohols are next; the displacement strength increases with the number of methylene groups. Methanol is weaker than ethanol, which in turn is weaker than isopropanol; acetonitrile is generally stronger than methanol, and its polarity is closer to propanol.

For chiral recognition, solvent strength is independent in most cases, since it affects only displacement of the analyte from the cavity. Methanol is commonly used as the starting solvent because it is a weak displacer. Stronger displacement effects are exhibited by acetonitrile, and it generally produces more efficient separation of peaks. For example, with ibuprofen, methanol overwhelms the weak hydrogen bonding forces of the secondary hydroxyl groups, diminishing separation, whereas acetonitrile produces the desired separation. Figure 10.8 details the operation of these columns in a reversed-phase mode.

It is necessary to strike a balance between the concentration of the buffer and the concentration of the organic modifier. Lower concentrations of the buffer (0.1% to 0.5%) allow incorporation of higher concentrations of organic modifier.

C. *Functional Group Interactions.* Certain functional groups have a strong affinity for the cyclodextrin cavity: iodide > bromide > chloride > fluoride; nitrate, sulfate, phosphate; and hydroxyls.

Other polar groups that can strongly hydrogen bond to the high-density diol surface are carboxyl and carbonyl; the amines (primary, secondary, or tertiary), free or in ring structures.

For chiral recognition to occur in the reversed-phase mode, the orientation of the analyte in the cyclodextrin cavity is quite important. Knowing the attraction of various groups to the cavity helps in these determinations. The retention of an aromatic structure can be increased by substitution with a preferred inclusion functional group, as well as by the presence of a preferred hydrogen bonding group.

1. To obtain chiral separation, the analyte must have a proper structural fit in the cyclodextrin. This generally dictates that at least one aromatic ring structure must be present in the solute molecule.
2. A substituent on or near the stereogenic center must interact (attract or repulse) with the 2- or 3-position hydroxyl groups at the mouth of cyclodextrin cavity. Steric effects due to large bulky groups near the stereogenic center can have a similar effect. It may be good to recall that these hydrogens are sterically fixed and the position of the interacting groups in relation to the aromatic ring may indicate better use of a derivatized cyclodextrin.

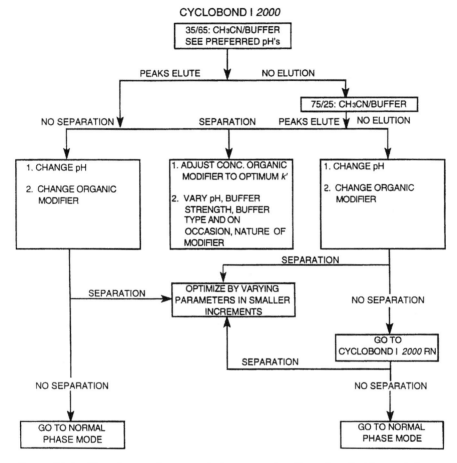

Figure 10.8 Flow diagram for reversed-phase operation. Reproduced with permission from *Chiral Separations*, Advanced Separations Technologies, Whippany, NJ. Copyright Advanced Separations Technologies, Inc.

D. *Operational Parameters.* The following parameters can influence separations: cyclodextrin phase, pH, buffer, flow rate, and temperature.

Cyclodextrin phase. The choice of a given cyclodextrin phase should be based on ring size, substitution position, and the distance bonding strength of the functional group in or near the stereogenic center. This information can be deduced from the polyaromatic hydrocarbon retention and preferred inclusion/hydrogen bonding groups as discussed.

Effect of pH. The obtained resolution is affected by the type of buffer, pH, and buffer concentration. To study the effects of each of these factors, an aqueous solution of 0.01M acetic acid is used and pH is adjusted by using dilute sodium hydroxide. For amines, pH is varied between 3 and 5, and for acids between 5 and 7; the effect on resolution is noted. A plot of resolution versus pH will help identify the proper pH for optimum resolution. The optimum pH thus selected is then used with a select type of buffer, and the effect of buffer concentration is then investigated. It is possible to control the hydrogen bonding of polar groups on or near the stereogenic center with the use of small amounts of weak acid and base.

Buffer effects. Buffers can also be included in the cyclodextrin cavity. As the buffer concentration is increased, solute peaks become sharper and the retention time decreases. For dansyl phenyl alanine, buffer concentration up to 1% can be used.[7] However, for ibuprofen (single aromatic ring), more dilute solutions are required. To reduce solute retention in the case of strongly retained analytes (for example with a naphthyl ring), ammonium nitrate can be used because it strongly hydrogen bonds to the primary hydroxyl groups. The preferred buffers are TEAA/TEA phosphate, citrate, ammonium nitrate, and ammonium acetate. The buffer concentration is generally in the mM range. When necessary, modifiers such as DEA, TEA, or HOAc are used in low concentrations.

Flow rate. The relationship of flow rate to efficiency is identical to reversed-phase HPLC if displacement from the cyclodextrin cavity is the only separation mechanism involved. However, for chiral compounds, the response follows a curve with a dramatic slope increase between 1.0 and 0.2 mL/minute. This has been observed for a wide variety of enantiomers.

Temperature effects. Low temperatures generally lead to increased retention and increased resolution. The degree to which temperature affects resolution is dependent on the analyte. To evaluate the effect of temperature, it is necessary to determine both α and N at three different temperatures — say, 5°C, 15°C, and 25°C. A plot of these values versus $1/T$ will be helpful in determining optimal conditions for chiral separations. In mobile-phase composition of 40% to 60% aqueous methanol, the higher viscosity at a lower temperature negates any beneficial effects. It may be desirable to use acetonitrile as the organic modifier in many cases.

2. Normal-Phase Separations. Normal-phase separations on cyclodextrin columns have generally been carried out with mobile phases such as hexane/isopropyl alcohol, acetonitrile/methanol, methanol, or ethanol. Researchers have found that π-π hydrogen bonding forces influence enantiomeric separations primarily. It was recently discovered that it is possible to override inclusion complexation in favor of interacting directly with secondary hydroxyl groups across the larger opening of a cyclodextrin toroid or the appendant carbamate, acetate, or hydroxypropyl functional groups. To accomplish this, polar organic mobile phases that produce very efficient separations can be used. The separations are generally more efficient and reproducible, retention time is frequently lower, and higher sample loads are possible. Beta-blockers such as propanolol, timolol, and atenolol, and compounds like

warfarin can be separated. The method also works for molecules that do not contain an aromatic group.

The surface of cyclodextrins is chiral and forms selective chiral interactions (diastereomers) with the hydrogen bonding groups of the analytes because the secondary alcohols of native cyclodextrins radiate from stereogenic centers. Derivatized cyclodextrins with acetate, naphthylethyl carbamate, or hydroxypropyl groups offer this potential as well, with the last two derivatives adding additional stereogenic centers. The uses of acid/base under anhydrous solvent conditions appears to affect the interactions of compounds with functional groups like alcohols, amines, carboxyls, and carbonyl group. Since three points of interactions are considered necessary for chiral recognition, the solvent pool in the cyclodextrin cavity can be seen as one of the interactions. At least two functional groups must be present in the analyte for this technique to work, and one of the interactive groups must be on or near the stereogenic center of the analyte.

The starting mobile phase may have the following components: acetonitrile, methanol, acetic acid, and triethyl amine. The mobile phase may be composed largely of acetonitrile (approximately 95%) with about 5% methanol.[7] The last two components are modifiers, and their amount is generally less than 0.3 %. The combined acid/base range can vary between 0.002% and 2.5%, with the average around 0.5%. To obtain optimum resolution, it is necessary to adjust only the ratio of these two components; the ratio can be 1:1 to 4:1, the typical average ratio of acid to base being 1.5:1. For carboxyl compounds, DEA has been at times substituted to increase the efficiency of the peaks.

An increase in retention time is obtained by reducing or eliminating methanol and decreasing acid/base concentration, while keeping the ratio constant. To decrease retention, the opposite has to be practiced, with methanol concentration and acid/base concentration increasing at the same ratio. There is very little effect of flow rates on resolution; however, it can help reduce retention time. The optimal flow rate is between 1.0 and 2.0 mL/minute.

At operating temperatures from ambient up to 40°C, little effect is seen on resolution, but the retention times can be reduced. To improve resolution, it is necessary to operate below the ambient temperatures. Figure 10.9 shows the operation of polar organic mode separations (normal phase).

Bonded Derivatized Cyclodextrins. Currently available derivatives are shown in Figure 10.10. The carbamate coupling of the π bases, 1-naphthylethyl to a bonded cyclodextrin, creates a complex environment that has demonstrated diverse chiral separations. It has been labeled a multimodal chiral stationary phase because it can be used in a normal- as well as a reversed-phase mode with appropriate modifiers. Analyte structure, solubility, and stability dictate the proper selection of the mobile phase. For example, if the analyte is π acidic, normal-phase solvents can be used. If the analyte is not π acidic but contains two hydrogen bonding groups, one on or near the stereogenic center, polar modifiers have to be used. Of the three carbamates available, the S-naphthylethyl carbamate (SN) has shown the greatest se-

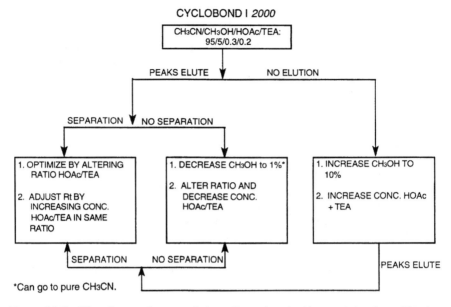

Figure 10.9 Flow diagram for normal phase. Reproduced with permission from *Chiral Separations*, Advanced Separations Technologies, Whippany, NJ. Copyright Advanced Separations Technologies, Inc.

lectivity and versatility. Since the naphthylethyl carbamate does play a role in enantioselectivity, the R form can be useful if the separation does not occur on the S form. Table 10.3 lists a variety of separations that have been obtained on cyclodextrin columns.

Macrocyclic antibiotics. Separations on cyclodextrin columns have been extended by utilizing macrocyclic antibiotics as chiral stationary phases by Armstrong et al.[8–10] These columns are sold as chirobiotic columns by Astec. Chirobiotic V HPLC is based on covalently bonding the amphoteric glycopeptide vancomycin to 5 μm silica gel. These ligands are linked to assure their stability while retaining essential components for chiral interactions. Vancomycin contains 18 chiral centers surrounding three pockets or cavities. Five aromatic ring structures bridge these cavities. Hydrogen donor and acceptor sites are readily available close to the ring structures. It has been claimed that the selectivity of this column is similar to glycoprotein AGP and it is stable when 0% to 100% organic modifier is used.

Chirobiotic T is based on bonding the amphoteric glycoside teicoplanin to 5 μm silica gel through covalent linkage. Teicoplanin contains 20 chiral centers surrounding four pockets or cavities. Hydrogen donor and acceptor sites are readily available close to seven aromatic rings.

Analytes. Neutral molecules, amides, acids, esters, and cyclic amines show considerable enantioselectivity. Other amines have been separated with

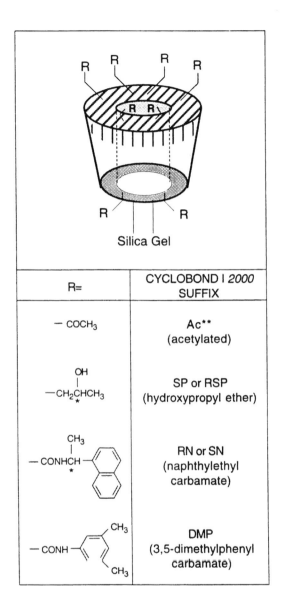

*Stereogenic Center.

**Note: Acetylated versions also available in gamma (CYCLOBOND II AC) and alpha (CYCLOBOND III AC).

Figure 10.10 Derivatized cyclodextrins. Reproduced with permission from *Chiral Separations*, Advanced Separations Technologies, Whippany, NJ. Copyright Advanced Separations Technologies, Inc.

Table 10.3 Separations on Cyclodextrin Phases

Compound	Column	α	Mobile Phase[a]
Catelol	β-CD	1.09	98:2:1.5:1
Labetolol	β-CD	>1 for 4 peaks	98:2:0.8:0.6
Nadolol	β-CD	>1 for 4 peaks	98:2:0.8:0.6
Metoprolol	β-CD	1.21	95:5:0.2:0.2
Pindolol	γ-CD	1.10	99:1:0.2:0.1
Propanolol	β-CD	1.12	95:5:0.3:0.2
Timolol	β-CD	1.15	95:5:0.3:0.2
Oxazepam	β-CD-RN	1.28	99:1:0.0064:0.0064
Suprofen	β-CD-RN	1.16	95:5:0.2:0.2
Coumachlor	β-CD-RN	1.25	98:2:0.8:0.6
Warfarin	β-CD-SN	1.23	95:5:0.005:0.003

a. Mobile phase components: ACN/MeOH/HOAc/TEA
Column; 250 × 4.6 mm; flow rate: 1 mL/min.; detector: UV 254 nm
RN-R-naphthylethylcarbamate; SN-S-naphthylethylcarbamate
Source: Reproduced with permission from reference 7. Copyright Advanced Separations Technologies, Inc.

varying degrees of success. A wide range of amino acid derivatives have been resolved on these columns.

Mechanism. Vancomycin has demonstrated broad selectivity in both reversed-phase and normal-phase solvents and limited selectivity with polar organic mobile phases. Since vancomycin contains peptide, carbohydrate, and other ionizable groups, it would be expected to offer different selectivity in these modes. The structure of vancomycin indicates that all of the typical interactions outlined for protein phases and other cellulosic polymer-type phases are possible with this phase. The potential interactions and their relative strengths are given below.

- Hydrogen bonding — very strong
- π-π Interactions — very strong
- Inclusion — weak because of shallow pockets
- Dipole stacking — medium-strong
- Steric interactions — weak
- Anionic or cationic binding — strong

Compared to cyclodextrins, the shallow pockets for inclusion yield weaker energies. This leads to faster kinetics, which can in turn lead to faster separations. Reversed-phase conditions favor inclusion and hydrogen bonding. Under these conditions, changes in pH can produce cationic or anionic interactions. Dipole stacking and π-π complexation are favored in normal-phase solvents. Polar organic mobile phases enhance the potential for all of these interactions.

The mobile phase functions equally in reversed-phase or normal-phase sollents because of the complex structure of the macrolide and ionizable groups. It does not function as well with the polar organic mobile phases as cyclodextrins do.

1. Reversed-Phase Separations. Optimization is accomplished by controlling the type and amount of organic modifier, type of buffer, and pH. Efficiency and selectivity are affected by ionic strength, buffer type, flow rate, and temperature. Of the various organic modifiers, tetrahydrofuran gives better selectivity and efficiency. With the reversed phases, higher aqueous composition is required than with C-18 columns. Typical composition of organic modifier/buffer (pH 4.0 to 7.0) is in the ratio of 10:90. Alcohols as modifiers generally require higher concentration—for example, 20% for comparable retention to ACN or THF. For chirobiotic T, a 20:80 ratio is a good starting point. The property order is methanol > ethanol > tetrahydrofuran > acetonitrile > isopropanol. For amino acids, ethanol exhibits higher selectivity.

Ammonium nitrate and triethylamine acetate buffers have been found useful. The optimum pH for resolution of various compounds is 4.0 to 7.0. For chirobiotic T, the pH range is 3.8 to 6.5. A decrease in pH to 3.8 produces a significant increase in retention of analytes within free carboxyl groups. In general, analytes act more favorably at a pH where they are not ionized. Thus the retention and selectivity of molecules that possess acidic or basic groups can be affected by changing the pH. A change in column temperatures has a significant effect. Lower temperatures are favored because of enhancement of weaker bonding forces.

2. Normal-Phase Separations. Peak efficiency and resolution can be improved with ethanol as the polar modifier of hexane. A good starting point may be 20% ethanol. In most cases, ethanol works better than isopropanol. The comparability of separations on Chirobiotic T and V for warfarin and N-CBZ-Norvaline are shown in Figures 10.11 and 10.12[11]

Table 10.4 provides some of the other applications of chirobiotic columns.

Polysaccharide columns (Type 2). The resolving capacity of polysaccharides such as cellulose was first realized with the observation that a racemic amino acid could occasionally give two spots in paper chromatography. This led Dalgliesh[12] to advance his three-point interaction theory in 1952 on the basis of results using paper chromatography of racemic amino acids. Optical resolution of amino acids on cellulose by TLC was reported. This has led to the use of cellulose and cellulose derivatives, as well as starch and cyclodextrins, for chiral separations by liquid chromatography. Cellulose, a linear polymer, has the chemical constitution of a linear poly-β-D-1,4-glucoside (see Figure 4.3).

Cellulose forms long chains at least 1,500 (+)-D-glucose units per molecule. The molecular weight of cellulose ranges from 2.5×10^5 to 1×10^6 or more. The (+)-D-glucose repetitive units contain five chiral centers and three hydroxyl groups. All the ring substituents are equatorial.

Broadly substituted cellulose columns can be divided into two major categories, as shown here:

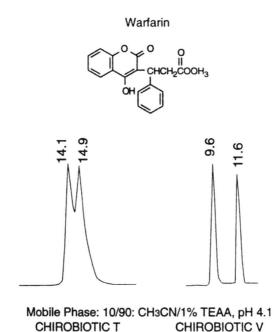

Figure 10.11 Comparability of separations on Chirobiotic columns. Reproduced with permission from *Chirobiotic Handbook*, Advanced Separations Technologies, Whippany, NJ. Copyright Advanced Separations Technologies, Inc.

Cellulose Esters	Cellulose Carbamates
Chiralcel CA-I	Chiralcel OC
Chiralcel OA	Chiralcel OD
Chiralcel OB	Chiralcel OD-H
Chiralcel OB-H	Chiralcel OD-R
Chiralcel OJ	Chiralcel OF
Chiralcel OK	Chiralcel OG

Table 10.5 provides a number of applications of these columns. Figure 10.13[13] provides a guide to selection of columns based on functionalities of the solute. The compounds are divided into four major categories based on whether they contain a certain functionality: aromatic, carbonyl, tertiary nitrogen, or hydroxyl.

Aromatic compounds containing carbonyl groups and no hydroxyls or tertiary nitrogen are better separated on Chiracel OJ, OD, and OC columns based on practical experience with a large number of compounds. If they were to contain a hydroxyl group as well and no tertiary nitrogen, then the choice is increased to four columns, where Chiracel OB and OT replace OD and OC. If, by contrast, tertiary nitrogen is also present, the choice is again narrowed to three columns where

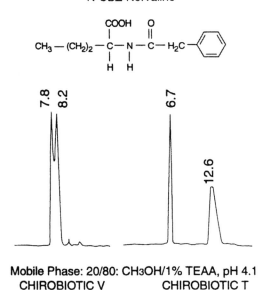

Figure 10.12 Comparability of separations on Chirobiotic columns. Reproduced with permission from *Chirobiotic Handbook*, Advanced Separations Technologies, Whippany, NJ. Copyright Advanced Separations Technologies, Inc.

Table 10.4 Applications of Chirobiotic Columns

Compound	Column[a]	Peak 1/Peak 2	Mobile Phase
Benoxaprofen	V	13.0/15.2 min.	100 THF:0.1% TEAA, pH 7.0
Ibuprofen	V	5.77/6.47 min.	100 THF:90 20 mM Na citrate, pH 6.3
Fenoterol	V	6.9/8.2 min.	10 THF:90 20 mM NH_4NO_3 pH 5.5
Mephobarbital	V	7.2/8.3 min.	25 EtOH:75 Hexane
Naproxen	V	8.8/10.5 min.	100 THF:0.1% TEAA, pH 7
Warfarin	V	9.6/11.6 min.	20 THF:80 0.1% TEAA, pH 5
Albuterol	T	10.1/11.4 min.	70 ACN:30 MeOH/ 0.5 HOAc/0.2 TEA
Citrulline	T	4.3/4.9 min.	40 MeOH:60 H_2O
DOPA	T	3.9/5.2 min.	50 EtOH:50 1% TEAA pH 4.5
Phenylalanine	T	6.2/9.2 min.	80 EtOH:20 H_2O

Source: Reproduced with permission from *Chirobiotic Handbook*, Advanced Separation Technologies, Whippany, NJ. Copyright Advanced Separations Technologies, Inc.

a. T = Chirobiotic T, V = Chirobiotic V.

Table 10.5 Applications on Polysaccharide Columns

Chiral Stationary Phase	Applications
Chiralcel OA	
Chiralcel OB, OB-H	Compounds with aromatic carbonyl, nitro, sulfinyl, cyano, or hydroxyl groups
Chiralcel OC	
Chiralcel OK	
Chiralcel OD, OD-H	Propanolol and other β-adrenergic blocking agents (without derivatization); alprenolol, metaprolol; practolol; pindolol; atenolol compounds with aromatic, carbonyl, nitro, sulfinyl, cyano, or hydroxyl groups
Chiralcel OD-R	Those where reversed-phase is needed
Chiralcel OF	Dihydropyridines, diltiazem, perisoxal, oxazolam, disopyramide, acetylphenetholide
Chiralcel OG	Dihydropyridines, diltiazem, acetylphenetholide
Chiralcel OJ	CNS depressants, arylpropionic esters, arylpropionic derivatives, insecticides with phosphorous-sulfur groups, mianserine, methaqualone, triadimefon, chlormezanone, phenobarbital, ethotoin, glutethimide, surecide, EPN, naproanilide
Chiralcel CA-1	Compounds with aromatic, carbonyl, nitro, sulfinyl, cyano, or hydroxyl groups
Chiralpak AD	Bulky substituent at the chiral center (e.g., diphenyl or pyridylmethane derivations); phenyl group attached to chiral center (e.g., hydroxyphenethylamine derivatives, arylacetic acid derivatives); heterocyclic compounds having a chiral center on the ring
Chiralpak AS	Heterocyclic compounds possessing a chiral center on the ring (e.g., diazepine derivatives, β-lactams)
Chiralpak WH	val, ile, phe, pro, ser, thr, tyr, trp, his. Partial resolution: ala, leu, met,
Chiralpak WM	ser, thr, gin, asp, gly, lys HCl
Chiralpak WE	ala, val, leu, ile, pro, gin. Partial resolution: met, arg HCl, tyr, glu, lys HCl
Chiralpak MA(+)	Hydroxy carboxylic acids
Crownpak CR	ala, val, leu, phe, DOPA, met, phe, ser, thr, cys, tyr, asp, glu, ornithine, lys, arg, citruline, his, trp (not proline), amines and amino alcohols
Bakerbond DNBPG	Secondary benzyl alcohols, mandelic acid analogs, aryl phosphonates, α indanol and α tetralol analogs, propanolol and analogs, aryl sulfoxides, aryl phthalides, aryl succinimides, aryl hydantoins, bi-naphthol analogs, aryl acetamides, acyclic alcohols, nitrogen heterocycles
Bakerbond DNB Leu	Aromatic phthalides, benzodiazepinones, substituted pyrolidine esters, aromatic amides
Bakerbond α-Naphthyl Urea	Ibuprofen, naproxen, fenoprofen, amphetamine, amethylbenzylamine, phenyl glycine, phenyl leucine
Chiral AGP	Alprenolol, atenolol, metaprolol, varapamil nitrendipine, felodipine, isopropylideneglycerol-4-methylbenzoyl ester, hexobarbital, 2-phenylbutyric acid, warfarin, ibuprofen, ethotoin, disopyramide, bupivacaine, ketamine, mepivacaine
Chiral-HSA	Kyrenine, folinic acid, N-benzoyl-D-leucine, 2-phenylpropionic acid, oxazepam, ketoprofen, fenoprofen, ibuprofen, naproxen, carprofen
Chiralpak OT(+), Chiralpak OP(+)	Compounds possessing an aromatic group

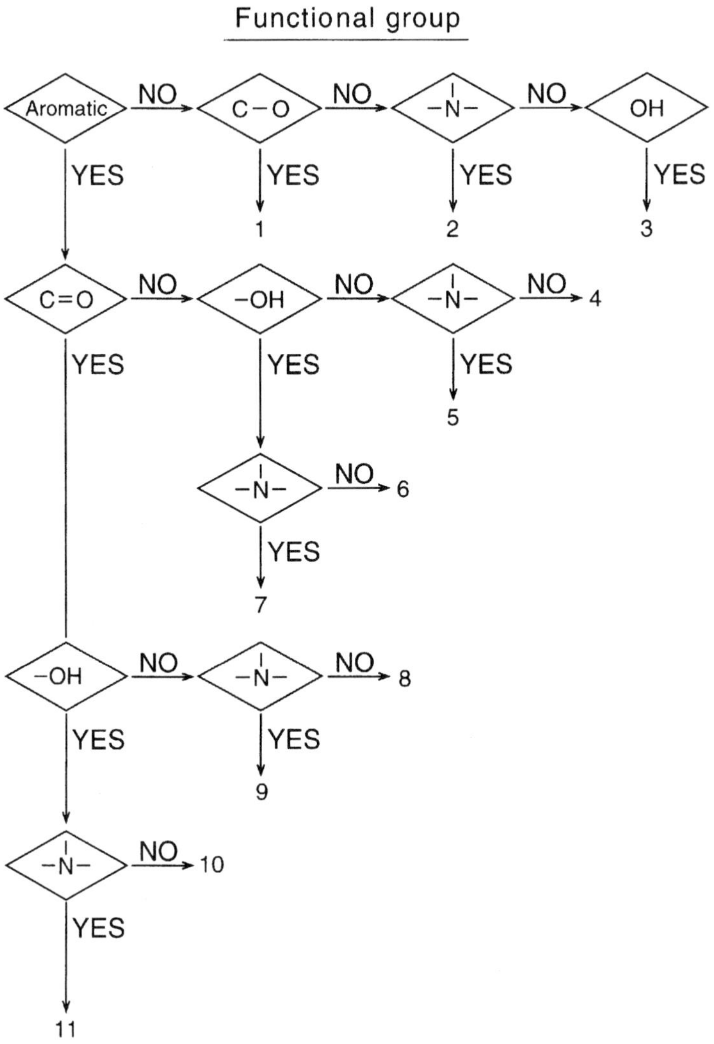

Functional group

1. Chiralcel OB, OD, AD, OJ, MA, WH
2. Chiralcel CR, WE, WH
3. Chiralcel AD, OB, OD, OJ
4. Chiralcel OD, OJ, OC
5. Chiralcel OC, OD, OJ
6. Chiralcel AD, OD, OJ, OB
7. Chiralcel CR, OJ, OD
8. Chiralcel OB, OJ, OD, OT
9. Chiralcel OB, OJ, OD, OC, OT, WH
10. Chiralcel OD, OJ
11. Chiralcel OC, OD, AD

Figure 10.13 Column selection based on group functionality. Reproduced with permission from *Application Guide for Chiral Column Selection*, Chiral Technologies, Exton, PA. Copyright Chiral Technologies.

Chiracel OJ is replaced by AD. It should be pointed out that information on selection of columns on this basis is somewhat empirical though based on a fairly large base of practical data.

Starch columns. Another polysaccharide built from (+)-D-glucose units is starch. It is composed of 20% amylose and 80% amylopectin, the latter being an insoluble fraction. Both are composed entirely of (+)-D-glucose units, linked by α-glucoside bonds. Amylose is a linear polymer; amylopectin is branched by C_1–C_6 connections.

Amylose columns. Chiralpak AD and Chiralpak AS are representatives of this type of column. Their applications can be found in Table 10.5.

Protein phases (Type 5). Chiral AGP is a second-generation chiral selector based on the α1-acid glycoprotein as the chiral stationary phase. The process of immobilizing AGP on porous spherical 5-μm silica has been patented. This CSP is useful for resolving a broad range of compounds such as racemic amines, acids, and nonprotolytic compounds without derivatization. A number of examples of drugs resolved on chiral AGP are given in Table 10.6.

The resolution ability of this column is due to the unique nature of the chiral stationary phase and the fact that enantioselectivity can be induced by choosing proper mobile phase composition. For bioanalytical work, this CSP has been highly recommended.

Table 10.6 Examples of Drugs Resolved on the Chiral AGP

Substance	Mobile Phase (modified)[a]	k_1	α
Alprenolol	7% 2-Propanol	17.90	1.19
Atenolol	Not modified	2.91	1.36
Bupivacaine	6% 2-Propanol	6.66	1.29
Chlorthalidone	5% Acetonitrile	3.68	2.45
Disopyramide	10% 2-Propanol	4.68	1.63
Ephedrine	5 mM Octanoic acid	3.86	1.34
Ethotoin	1.5% 2-Propanol	0.99	2.45
Felodipine	12.5% 2-Propanol	11.40	1.34
Fenoprofen	1% 2-Propanol (pH 6.5)	6.88	1.29
Hexobarbital	3% 2-Propanol	1.26	1.59
Ketamine	2.5% 2-Propanol	7.00	1.26
Metoprolol	0.5% 2-Propanol	4.72	1.26
Pheniramine	7% Acetonitrile	11.30	1.33
2-Phenoxypropionic acid	Phosphate buffer (pH 5.5)	1.91	1.57
Terodiline	15% 2-Propanol	12.40	1.33
Tiaprofenic acid	1% 1-Propanol (pH 6.5)	4.16	1.75
Tropicamide	3% Acetonitrile	9.09	1.38
Verapamil	10% Acetonitrile	15.60	1.32

a. Phosphate buffer (10 mM) at pH 7.0 as modified (Source: G. Schill).

The typical mobile phase for this column is phosphate buffer with an organic modifier. Enantioselectivity and retention can be regulated by changing the mobile-phase composition with respect to any of the following parameters: pH, organic modifier, modifier concentration. For the various types of modifiers that have been used and their respective concentrations, see Table 10.6. Typical operating conditions entail using 10 mM buffer at pH 7.0 with or without organic modifier. Modifier concentrations as high as 15% isopropanol or 10% acetonitrile have been used.

Effect of pH. The effect of pH on resolution of different analytes is given in Table 10.7. The four compounds selected are 2-phenoxypropionic acid, acidic; ethotoin, nonprotolytic; metoprolol, amine; and hexobarbital, weakly acidic. For chromatographing hydrophobic amines, a pH of 4 to 5 is preferred. In this pH range, the protein has a lower negative charge than at pH 7. This leads to lower affinity of amines for the column, resulting in lower retention time.

Buffer concentration. It is possible to affect both retention and enantioselectivity by changing the buffer concentration of the mobile phase. This is shown in a plot of buffer concentration and k' for ibuprofen (Figure 10.14).

Modifier Concentration. In general, increasing the concentration of the modifier will decrease retention and may improve enantioselectivity (Table 10.8). However, for some acidic compounds, a different behavior is observed; increasing the modifier concentration results in decreasing retention and an increase in enantioselectivity. This type of behavior is exemplified by warfarin (Table 10.9).

Nature of modifier. It is possible to obtain different enantioselectivity by changing one organic modifier to another with different hydrogen bonding properties and different hydrophobicity. For example, with some compounds such as pindolol, no resolution is obtained with 1-propanol as the modifier; however, baseline resolution is obtained with acetonitrile.

Method development. A separate scheme for method development for the following classes of compounds is given in Figures 10.15[14] to 10.17. It

Table 10.7 Effect of pH on Resolution (α values)

Compound	pH 4.5	pH 5.5	pH 6.5	pH 7.5
2-Phenoxypropionic acid	1.59	1.57	1.48	—
Ethotoin	3.82	4.19	4.59	5.06
Metoprolol	1.25	1.29	1.42	1.48
Hexobarbital	1.44	1.47	1.66	2.10

Note: Mobile phase: 0.01 M phosphate buffer

Source: Reproduced with permission from ChromTech Application Note no. 13, 1994. Copyright 1994 Chrom Tech.

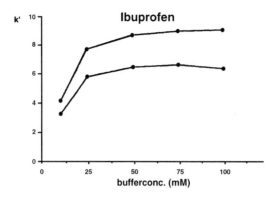

Figure 10.14 Effect of buffer concentration on separation of ibuprofen. Reproduced with permission from ChromTech Application Note no. 13, 1994. Copyright 1994 ChromTech.

should be pointed out that these schemes are based on practical experience of various users, which can be helpful in designing a method to fit individual needs.

Human serum albumin (HSA) has been immobilized onto 5-µm spherical silica particles. The surface chemistry of silica and the method of immobilization provide a stable chiral separation material. Both racemic acids and amino acids can be resolved directly without derivatization. Enantioselectivity and retention can be regulated by changing the mobile-phase composition, pH, buffer concentration, and/or the nature of the organic modifier.

The column is operated in the reversed-phase mode. A suitable phosphate buffer is utilized (0.01 to 0.1 M, pH 5 to 7), with the addition of less than 10% or-

Table 10.8 Influence of Modifier Concentration on Retention and Enantioselectivity

2-Propanol Concentration (%)	k_1	k_2	α	RS
Metoprolol solute				
0.5	9.71	12.3	1.27	1.66
2.5	3.97	4.38	1.10	1.10
5	2.56	2.56	1.00	—
Methylphenylcyanoacetic acid ethyl ester solute				
5	2.53	4.18	1.65	3.64
7.5	1.93	2.99	1.55	3.13
10	1.42	2.02	1.42	2.20

Note: Mobile phase: modifier in 10 mM sodium phosphate buffer (pH 7.0).

Table 10.9 Influence of 2-Propanol Concentration on Retention and Enantioselectivity of Warfarin

Concentration (%)	k_1	α
8	4.73	1.33
10	2.45	1.42
12	1.19	1.53
14	0.76	1.57

Note: Mobile phase: 2-propanol in 0.01 M phosphate buffer (pH 7.0).

ganic modifier such as 2-propanol, acetonitrile, methanol, or ethanol. Charged modifiers such as octanoic acid (1–5 μm) may also be used.

Ligand exchange columns (Type 4). A number of ligand-exchange columns are sold by Daicel, including Chiralpak WH, Chiralpak WM, and Chiralpak WE. These columns are useful for resolution of amino acids and their derivatives.[15] The mobile phase is simply aqueous copper sulfate. Chiralpak Ma (+) is another ligand-exchange (coating type) column that is useful for hydroxycarboxylic acids.

Davankov and Kurganov[16] were the first to indicate that cross-linked resins with fixed ligands, (R)-N,N'-dibenzyl-1,2-propanediamine in the form of copper(II) complexes, display high enantioselectivity in LEC of amino acids such as lysine, alanine, serine, and leucine. Various amino acids, including baclofen (**10.1**), can be resolved on a reversed-phase C-18 column with a chiral mobile phase of aqueous cupric acetate and N,N-di-n-propyl-L-alanine (DPA) containing 15% acetonitrile.[17,18]

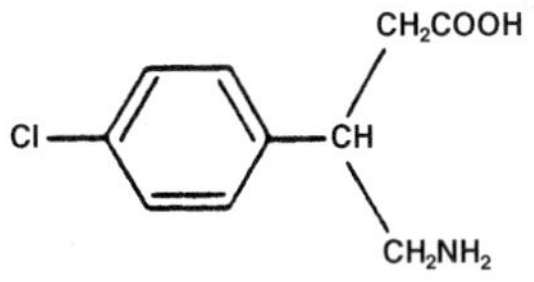

BACLOFEN
(β-chlorophenylGABA)

10.1

Cation-exchange chromatography can then be used to break the Cu-DPA-baclofen complex on a Dowex-50W column to yield small quantities of the optical isomers for mechanistic studies.

Investigations have revealed that a ligand-exchange column may be a simpler and more useful approach for separating enantiomers of baclofen. A Chiralpak WH column was selected for this purpose, since this column was designed to serve as ligand exchanger.[19] Unfortunately, the nature of its stationary phase is proprietary.

It is fairly well known that the separation of amino acids is significantly affected

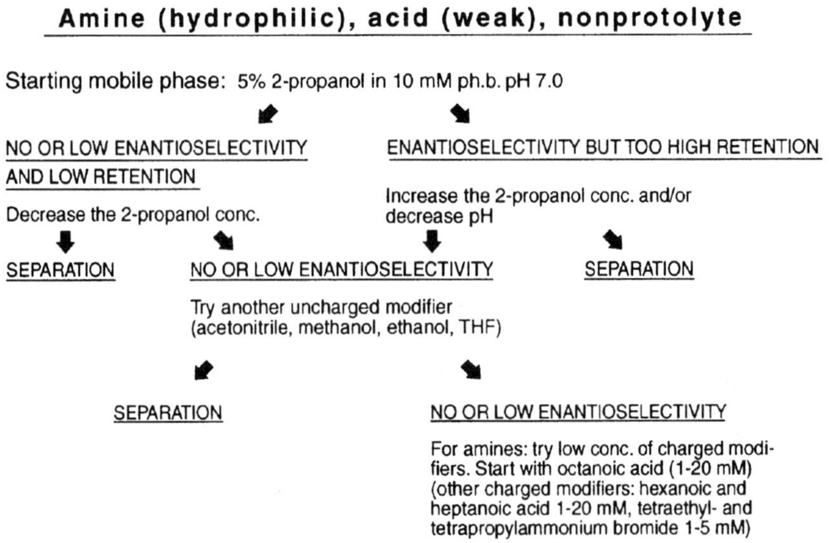

Amine (hydrophilic), acid (weak), nonprotolyte

Starting mobile phase: 5% 2-propanol in 10 mM ph.b. pH 7.0

- ↙ NO OR LOW ENANTIOSELECTIVITY AND LOW RETENTION
  Decrease the 2-propanol conc.
  - ↓ SEPARATION
  - ↘ NO OR LOW ENANTIOSELECTIVITY
    Try another uncharged modifier (acetonitrile, methanol, ethanol, THF)
    - ↙ SEPARATION
    - ↘ NO OR LOW ENANTIOSELECTIVITY
      For amines: try low conc. of charged modifiers. Start with octanoic acid (1-20 mM) (other charged modifiers: hexanoic and heptanoic acid 1-20 mM, tetraethyl- and tetrapropylammonium bromide 1-5 mM)
- ↘ ENANTIOSELECTIVITY BUT TOO HIGH RETENTION
  Increase the 2-propanol conc. and/or decrease pH
  - ↓ SEPARATION

Figure 10.15 Separations of hydrophilic amines, weak acids, and nonprotolytic compounds. Reproduced with permission from ChromTech Application Note no. 13, 1994. Copyright 1994 ChromTech.

by the molarity of $CuSO_4$, which was used for the separation (Figure 10.18). However, the effect of organic solvents such as methanol on the retention of free amino acids is not well known, except for valine, which, with an increase in concentration of methanol up to 20%, shows a decrease in retention time. Temperature has a significant effect on retention; for example, the retention time of phenylala-

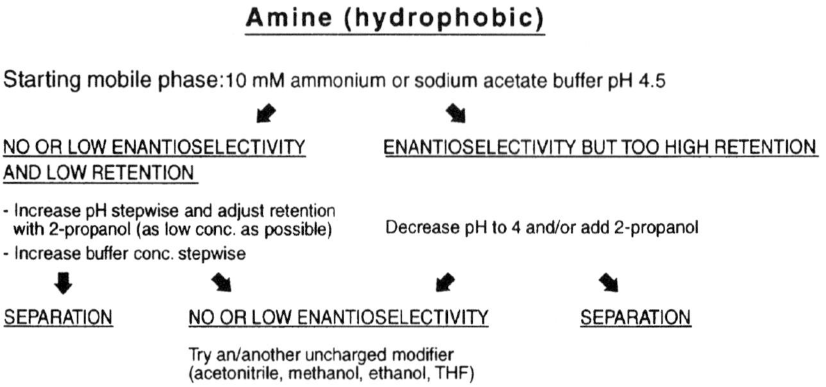

Amine (hydrophobic)

Starting mobile phase: 10 mM ammonium or sodium acetate buffer pH 4.5

- ↙ NO OR LOW ENANTIOSELECTIVITY AND LOW RETENTION
  - Increase pH stepwise and adjust retention with 2-propanol (as low conc. as possible)
  - Increase buffer conc. stepwise
  - ↓ SEPARATION
  - ↘ NO OR LOW ENANTIOSELECTIVITY
    Try an/another uncharged modifier (acetonitrile, methanol, ethanol, THF)
- ↘ ENANTIOSELECTIVITY BUT TOO HIGH RETENTION
  Decrease pH to 4 and/or add 2-propanol
  - ↓ SEPARATION

Figure 10.16 Separation of hydrophobic amines. Reproduced with permission from ChromTech Application Note no. 13, 1994. Copyright 1994 ChromTech.

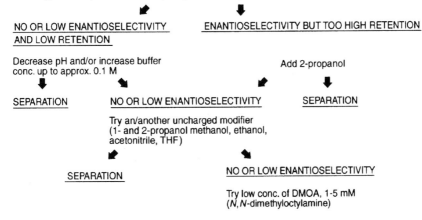

Figure 10.17 Separation of strong acids. Reproduced with permission from ChromTech Application Note no. 13, 1994. Copyright 1994 ChromTech.

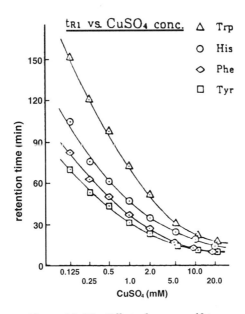

Figure 10.18 Effect of copper sulfate.

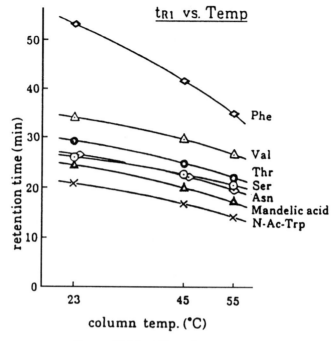

Figure 10.19 Effect of temperature.

nine decreases significantly with increasing temperatures. Other amino acids do not show as great a change, as indicated in Figure 10.19.

Based on these considerations, several mobile phases were investigated. No advantage was seen in using methanol in the mobile phase. The mobile phase containing 0.25 mM copper sulfate, run at a flow rate of 1.5 mL/minute at 50°C, gave the optimal resolution of d- and l-forms of baclofen. The retention times of the two enantiomers separated from the racemate was 4.75 and 5.43 minutes. The enantiomer with lower retention time was identified as the l-form, based on the injections of authentic l-baclofen and d-baclofen.

Miscellaneous Columns

Crown ether columns. Macrocyclic polyethers are called crown ethers because molecular models of them resemble crowns. They are well known for forming host-guest complexes with metal cations and substituted ammonium ions. Chiral crown ether, as a chiral selector, is useful for compounds such as those possessing primary amino groups near asymmetric centers, because the multiple hydrogen bonds formed between the ammonium ion ($-NH_3^+$) produced under acidic conditions and the ether oxygens, for steric reasons, lead to a less stable complex with one of the enantiomers in the chiral cavity of the crown ether (10.2).

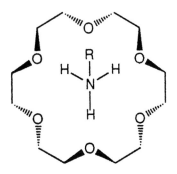

Crown Ether CSPs

10.2

Crown ether columns are available from Daicel as follows: Crownpak CR(+) and Crownpak CR(−). These columns resolve compounds possessing primarily amino groups such as amino acids, amino alcohols, and simple amines. Good resolution is possible even for the compounds poorly resolvable by conventional ligand exchange columns (LECs). UV detection is greater than by LEC requiring $Cu_2{}^+$ in the eluent.

The CR(+) column requires acidic conditions so that the primary amino group of a sample can be converted to an ammonium ion. Although various acids can be used for this purpose, perchloric acid is a good choice because it does not interfere with UV monitoring and helps increase the retention on the column. A shorter retention time is observed with greater hydrophilicity of the analyte. A decrease in temperature and/or pH can help enhance sample retention. A number of amines, amino alcohols, and amino acids have been resolved on these columns. Separations of amino acids are given in Table 10.10.

Polymethacrylate columns. The following polymer-based columns are available from Daicel: Chiralpak OT(+) and Chiralpak OP(+). They are useful for compounds with aromatic groups. Simple mobile phases such as methanol or hexane/2-propanol can be used.

Additional Applications Based on Structure of Analyte

A fairly large number of applications that have already been discussed under various column headings are not being duplicated here. Additional applications have been classified into the following groups shown here and on the next page, based on structure, to further assist the readers in matching their applications to those available in the literature.

Alcohols/hydroxy compounds	Hydantoins
Alkaloids/basic nitrogen compounds	Benzothiadiazines

Amides/imides
Amines
Amino acids
Barbiturates
Carboxylic compounds

Benzodiazepines
Esters
Phosphorous compounds
Sulfur compounds
Miscellaneous compounds

Alcohols/hydroxy compounds. A number of hydroxy compounds such as *trans*-2-phenyl cyclohexanol, tetrahydrobenzopyrene-7-ol, ibuprofenol, and tetrahydronaphthol have been resolved on Whelk-O 1 columns.[5] The mobile-phase composition included hexane and isopropanol at varying proportions ranging from 80% to 99%. The α values ranged from 1.05 to 1.18.

The enantiomers of compounds such as alprenolol and atenolol have been resolved on a Chiral AGP column (100 × 4.0 mm). The mobile phase for alprenolol was composed of 3% acetonitrile in 10 mM sodium acetate at pH 4.0, while the mobile phase for atenolol had no acetonitrile and consisted mainly of 10 mM

Table 10.10 Resolution of Amino Acids on Crownpak CR(+)

DL-Amino Acid	α	$HClO_4$/pH	Temperature (°C)
Alanine	1.86	1.5	25
Valine	1.51	1.5	0
Norvaline	1.69	2.0	25
Leucine	1.67	2.0	25
Norleucine	1.66	2.0	25
Isoleucine	1.58	2.0	0
tert-Leucine	1.10	1.0	0
Phenylalanine	1.27	2.0	25
Dopa	1.32	2.0	25
Methionine	2.00	2.0	25
Ethionine	1.93	2.0	25
Phenylglycine	2.35	2.0	40
Serine	1.75	1.5	0
Threonine	2.58	2.0	0
Cysteine	1.67	1.5	25
Tyrosine	1.28	2.0	25
Asparagine	1.69	1.5	0
Glutamine	2.13	2.0	25
Asparadic acid	2.01	2.0	0
Glutamic acid	2.81	2.0	25
Ornithine	1.49	1.5	25
Lysine	1.26	1.5	25
Arginine	2.21	1.5	25
Citruline	2.18	1.5	25
Proline	1.00	1.5	0
Histidine	1.82	1.5	0
Tryptophan	1.19	2.0	25

sodium acetate buffer at pH 7.0.[14] It should be pointed out that these compounds also contain substituted amine and phenyl groups.

Other examples of separations of compounds with amino and hydroxyl functions on AGP column are clenbuterol, ephedrine, epinephrine, salbutamol, and metoprolol. Metoprolol, which also has these functions, is better resolved on a Chiral-CBH column (100 × 4.0 mm). The mobile phase contains 5% 2-propanol in 10 mM sodium phosphate buffer at pH 6.0, plus 50 µM disodium EDTA.

Alkaloids/basic nitrogen compounds. Atropine and homatropine have been resolved on Chiralcel OD column with hexane:IPA:diethylamine (80:20:0.1) mobile phase at 0.5 mL/min.[13] Atropine can also be resolved on a Chiral AGP column (100 × 4.0 mm) with a mobile phase containing 2% 2-propanol and 5 mM octanoic acid in 0.01 M sodium phosphate buffer at pH 7.0.[14] Other alkaloids such as *Vinca* alkaloids have been resolved on Chiral AGP columns (50 × 4.0 mm) with 20% acetonitrile in 10 mM sodium phosphate buffer at pH 6.0.[14]

Nitrogen compounds such as Troeger's base can be resolved on a Whelk-O 1 column (4.6 mm i.d. × 25 cm). The mobile phase is composed of hexane and isopropanol in a ratio of 97:3. The enantiomers gave an α value of 1.4 at a flow rate of 1 mL/min.[5]

Amides/imides. A number of amides and imides have been resolved on Whelk-O 1 columns. Some of the notable examples of separations of amides are CBZ-Val or CBZ-Phe.[5] Hexane:IPA:HOAc (95:5:0.1) has been used as a mobile phase. For both of these separations, a 4.6 mm × 25 cm column is used at a flow rate of 1 mL/min, and detection is performed at 254 nm by UV. The run times are 20 and 40 minutes, respectively. For CBZ nornicotine (**10.3**), the mobile phase can be modified to methanol:dichloromethane to produce a shorter run time of 5 minutes.

10.3

Resolution of an imide with the structure as shown in 10.3 was achieved with a mobile phase containing hexane:IPA:HOAc (80:20:0.1). The α value for the enantiomers was 1.36.

The enantiomers of bupivacaine can be resolved on a chiral AGP column (100 × 4.0 mm) with 8% tetrahydrofuran in 10 mM sodium phosphate buffer at pH 7.0. Alternatively, 6% 2-propanol in 0.01 M phosphate buffer at pH 7.0 can be used as a mobile phase. It should be noted that this compound has a piperidine ring and an aromatic ring.[14]

Amines. Derivatization of primary amines has led to resolution of these compounds on a cyclodextrin column.[7] One notable example is the separation of enantiomers of amphetamine derivatized with Accu-Tag (Waters Corp.), followed by chromatography on Cyclobond phases (Astec). The best resolution was obtained on Cyclobond I 2000 DMP column (4.6 mm i.d. × 25 cm) with a mobile-phase composition of ACN: 0.1% TEA (45:55) at pH 4.1 and a flow rate of 1 mL/min. The L-isomer elutes first, and a resolution value of 2.4 is obtained.

Compounds such as metoprolol, practolol, pindolol, propanolol, and alprenolol have been resolved on Chiralcel OD columns.[7] These compounds are multifunctional and contain a substituted amine function, a hydroxyl group, and a phenyl ring. The mobile phase utilized was composed of hexane:IPA:diethylamine (80:20:0.1). Detection was carried out at 254 nm.

Amines and amino alcohols can be resolved on Crownpak CR(+) columns at pHs between 1.0 and 2.0 at ambient temperature. Low temperatures down to 1°C are frequently used to improve separations.[7]

Aliphatic amino compounds can be easily protonated to yield cationic compounds. A number of important drugs contain an aliphatic amino group. The notable examples are anticholinergic alkaloid derivatives, amino alcohols with β-blocking action, and a variety of local anesthetics. Propanolol (already discussed) is one such drug that has elicited significant interest. It has been chromatographed on Pirkle (R)-N-(3,5-dinitrobenzoyl)phenylglycine column.[20] The procedure requires derivatization with phosgene to produce its oxizolidinone, followed by HPLC with hexane:2-propanol:acetonitrile (96:3:1). Detection is performed fluorimetrically with excitation at 290 nm and emission at 335 nm. Propanolol can be detected in whole blood at 10 ng/mL level. The α values of oxizolidinones of propanolol is 1.09. Pronethalol (**10.4**) can also be resolved with this method.

10.4

Pindolol has been resolved by derivatization with isopropyl isocyanate on a Resolvosil BSA-silica column coupled to thermospray-MS system.[21] Direct resolution of β-blockers is possible by ion-pair chromatographic techniques with chiral counterions and the use of an AGP column.[22] Five β-blockers (alprenolol, oxyprenolol, propanolol, pindolol, and atenolol) have been completely resolved on silica-based cellulose tris (3,5-dimethyphenylcarbamate) columns with a mobile phase composed of hexane:2-propanol (9:1). Separation of these compounds is also possible using CBH-I columns.[23]

Amino acids. HPLC provides a significant advantage in separation of amino acids in that derivatization often is not required. Columns based on albumin have shown remarkable enantioselectivity toward N-acylated amino acids. Among the derivatives that can be resolved are those that possess highly fluorescent properties, such as dansyl and FTIC derivatives. Carbazole derivatives can be easily prepared by reacting amino acids with N-(chloroformyl)-carbazole in an acetone buffer mixture at pH 9. The derivatives of protein amino acids that can be resolved into enantiomers with high separation factors are given in Table 10.11.[24] All of these derivatives can be detected at lower concentrations than FMOC derivatives and are stable in dilute solutions. Table 10.12 gives separations of amino acids on chirobiotic columns. Separations on Crownpak CR column have been provided in Table 10.10.

Chiralpak WH columns have been used to resolve a number of amino acids, based on ligand exchange. Examples of such compounds are valine, serine, threonine, methionine, aspartic acid, phenylalanine, tyrosine, and histidine. No derivatization is required. The mobile phase contains 0.25 mM cupric sulfate, and separations are generally carried out at higher than ambient temperatures. For these separations, 50°C was found to be the preferable temperature.[12]

Barbiturates. These compounds have a pKa value of approximately 8.5 and are generally stable to racemization. Differences in hypnotic activity have been observed between the enantiomers of barbiturates. For example, it has been demonstrated that the (S)-(+)-enantiomer of hexobarbital (R,R, = R1 = CH_3, R2 = 1-cyclohexenyl) is more potent as a hypnotic than the (R)-(−)-form.

Hexobarbital, mephobarbital, and a number of other barbiturates (**10.5**) have been resolved on a number of CSPs such as microcrystalline cellulose triace-

Table 10.11 Some Useful Derivatives of Amino Acids for Optical Resolution by Chiral LC on BSA-Based Columns

	λ (nm)	Elution Order of ala-Derivative
UV Detection Derivative		
N-(4-nitrobenzoyl)-	269	L<D
N-(3,5-dinitrobenzoyl)-	269	L<D
N-(2,4-dinitrobenzoyl)-	340	L<D
N-(phthalimido)-	225	L>D
Fluorescence Detection Derivative		
N-(5-dimethylamino-1-napththalene-sulfonyl)-(DANSYL)	347/396	L>D
N-(9-fluorenylmethoxycarbonyl)-(FMOC)	260/310	L<D
N-(fluoresceinthiocarbamoyl)-(FITC)	488/520	n.d.

Source: Reproduced with permission from S. Allenmark and S. Anderson, *Chromatographia*, 31, 429 (1990). Copyright 1990 *Chromatographia*.

tate (MCTA), poly(triphenylmethylmethacrylate), cyclodextrins, and Pirkle-type columns.[25]

[structure 10.5: barbiturate ring with R_1, R_2, and R substituents]

10.5

Mephobarbital and hexobarbital[25] have been resolved on Daicel columns. For mephobarbital, Chiralcel OJ can be used with ethanol as the mobile phase at a flow rate of 1 mL/min. However, 5% water has to be added to ethanol to resolve hexobarbital on Chiralcel CA-I columns. A low flow rate is used, and detection is carried out at 230 nm instead of 254 nm for mephobarbital.[13] Hexobarbital can

Table 10.12 Chromatographic Parameters for the Enantioresolution of 20 Naturally Occurring Amino Acids

Amino Acid	k_1	k_2	α	R_S
Aspartic acid	0.20	0.34	1.7	1.2
Threonine	0.28	0.39	1.4	1.1
Glutamic acid	0.30	0.57	1.9	1.5
Serine	0.33	0.45	1.4	1.2
Isoleucine	0.40	0.80	2.0	2.5
Glutamine	0.40	0.72	1.8	1.6
Glycine	0.41	—[a]	—	—
Tyrosine	0.42	0.64	1.5	1.9
Cysteine	0.45	0.72	1.6	1.6
Valine	0.46	0.75	1.6	1.9
Leucine	0.48	1.01	2.1	3.5
Methionine	0.53	1.16	2.2	3.3
Phenylalanine	0.56	0.83	1.5	2.0
Alanine	0.56	1.03	1.8	2.9
Proline	0.58	1.46	2.5	2.5
Asparagine	0.60	0.98	1.6	2.1
Tryptophan	0.77	1.17	1.5	2.2
Lysine	6.12	9.18	1.5	2.2
Arginine	6.48	8.96	1.4	2.1
Histidine[b]	6.60	7.60	1.2	0.8

Note. Data was generated with a 250 × 4.6 mm Chirobiotic T (5-μm teicoplanin-bonded silica), methanol:water 60:40 mobile phase, 1 mL/min., 210 nm UV detection of underivatized solutes.

a. Achiral.
b. Mobile phase pH adjusted to 3.80 by acetic acid.

also be resolved on Chiral AGP columns (100 × 4.0 mm) with 2% 2-propanol in 10 mM sodium phosphate buffer at pH 7.0.[14] Other barbiturates that have been resolved on Chiral AGP columns are methylphenbarbital, penthiobarbital, secobarbital.

Carboxylic acids. Etodolac, fenoprofen, and flurbiprofen have been resolved on Chiral AGP columns with varying proportions of 2-propanol in sodium phosphate buffer at pH 6.5 to 7.0. One mm dimethyloctylamine helped improve the separations for flurbiprofen. Leucovorin required a Chiral HSA column with 6% 2-propanol in 100 mm sodium phosphate buffer at pH 7.0.[14] Ibuprofen can be resolved on Chiral AGP or Chiral HSA columns. The separations are achieved using sodium phosphate buffer at pH 7.0, with the addition of 1 to 5 mM of dimethyloctylamine or octanoic acid.[14] Other compounds that have been similarly resolved are ketoprofen and naproxen.

Analgesics based on 2-arylpropionic acid, such as ibuprofen, have been resolved on BSA-, AGP-, and OVM-based columns. Table 10.13 summarizes various techniques that have been used for chromatographing this group of drugs.[26]

Hydantoins. Mephenytoin can be resolved on a Chiral AGP column (100 × 4.0 mm) with 0.3% 1-propanol in 100 mM sodium phosphate buffer at pH 7.0.[14] Several hydantoins, including mephenytoin, can be resolved by HPLC on Pirkle columns.[27] Mephenytoin gives an α value of 1.06 on a covalent DNB-leu sorbent and hexane:2-propanol:acetonitrile (89:10:1).

Benzothiadiazines. Benzothiadiazines (**10.6**) are used as diuretics and have the following structure:

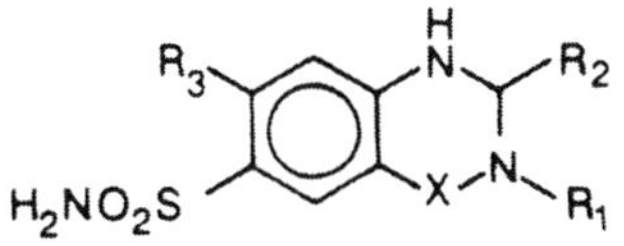

$X=SO_2$ (benzothiadiazines)

10.6

Optical resolution of a series of racemic benzothiadiazines has been achieved on polyacrylamide phases by LC.[28] These methods were used to obtain high enrichment of a given enantiomer. Semipreparative-scale separations with repetitive chromatography yielded some compounds with fairly high optical purity. For example, enantiomers of penflutizide and bendroflumethazide have been obtained at >97% optical purity and used for polarimetric studies for racemization kinetics. Pseudo first-order reaction rates have been observed in aqueous ethanol under slightly alkaline conditions. Racemization studies show that the half-life decreases with pH, being only a few minutes at pH >9. The base-cat-

Table 10.13 Techniques Used for Optical Resolution of Some Anti-inflammatory Agents of the α-Arylpropionic Acid Type

Compound	Derivative	Column/Mobile Phase
Ibuprofen	—	AGP/buffer
	Amide (with NMA)	PG/7.5% dioxan in hexane
Flurbiprofen	Amide (with NMA)	PG/3% 2-propanol in hexane
	Amide (with NMA)	CTPC/3% methanol in hexane
Benoxaprofen	Amide (with NMA)	PG/2-propanol in hexane
Fenoprofen	Amide (with NMA)	PG/2-propanol in hexane
Naproxen	Amide (with NMA)	PG/2-propanol in hexane
	—	BSA/buffer
Ketoprofen	—	BSA/buffer

Note: AGP = α_1-acid glycoprotein column (Enantiopac), BSA = silica-based bovine serum albumin column (Resolvosil), CTPC = cellulose tris(phenylcarbamate) coated onto silica (Chiralcel OC), NMA = 1-napthyl-methylamine, and PG = covalent (R)-N3,5-dinitrobenzoyl)phenylglycine (Hi-Chrom Reversible).

alyzed racemization proceeds simultaneously with the hydrolysis of thiadiazine observed previously.

Benzodiazepinones. The compounds in this class have been used as sedatives and hypnotics. The compounds are generally too polar for favorable GC separations; however, a number of them have been resolved by HPLC. For example, oxazepam was resolved on a preparative scale on a chiral polyacrylamide sorbent (ethyl [S]-phenylalaninate substituent). Hexane:dioxane (65:35) is used as the mobile phase when this polymer is deposited on silica as a stationary phase.[29] Another silica-based CSP based on N-formyl-(S)-phenylalanine has been found useful for analytical separations of enantiomers of temazepam and camazepam.

Pirkle and coworkers have resolved 42 benzodiazepinones on an N-(3,5-dinitrobenzoyl)-phenylglycine and leucine sorbents.[30] In all cases, the (−)-form elutes first on the (R)-phenylglycine column but last on the (S)-leucine column. The latter column gave the largest separation factor; the α values for these resolutions vary between 1.07 and 4.33.

The most stable conformation of a 3-substituted diazepam is one with the substituent occupying an equatorial position. This conformation tends to direct the amide carbonyl group of the folded ring toward the amide hydrogen atoms of the bound selector. The following bonding interactions are indicated:

1. Charge transfer interaction between the 3,5-dinitrobenzoyl group and the benzene ring of the analyte
2. Hydrogen bonding between the 3,5-dinitrobenzamide hydrogen and the middle carbonyl oxygen atom of the analyte
3. Hydrogen bonding between the amino acid carbonyl oxygen atom of the CSP and the amide hydrogen atom of the analyte

It appears that the length of the 3-substituent does not affect resolution, suggesting that it does not contribute to steric interactions. A number of compounds in this group can also be resolved on albumin silica columns.[31] The retention on these columns is highly dependent on the hydrophobic character of the 3-substituent, and the mobile phase has to be carefully selected to give reasonable k' values.

Esters. Various esters have been resolved on Whelk-O 1 columns, with a mobile phase containing hexane:IPA:HOC (95:5:0.5) with a 4.6 mm i.d. × 25 cm column.[5] The composition for methyl mandelate required water:ACN:HOAc (73:27:0.1) to produce an α value of 1.15.

Phosphorus compounds. A number of phosphorus enantiomeric compounds can be separated on Whelk-O 1 columns (4.6 mm i.d. × 25 cm). The mobile phase can be as simple as 10% ethanol in hexane. Isopropanol can replace ethanol for some separations.[5]

Sulfur compounds. A fairly large number of sulfoxides (**10.7**) have been resolved on the standard Whelk-O 1 column with a mobile phase containing hexane:IPA:-methylene chloride (7:2:1). A flow rate of 2 mL/min can be utilized for these separations, and detection is carried out at 254 nm.[5] Alpha values ranged from 1.1 to 1.5.

10.7

Ethiazide was resolved on a Chiralpak AAS column with a hexane:ethanol (80:20) mobile phase at a flow rate of 1 mL/min at 40°C. The detection was by UV at 254 nm. Good resolution was obtained in 45 minutes. Benzyl hydrochlorothiazide required the addition of 0.1% diethylamine, a competing amine, for the resolution of the enantiomers.[14]

Miscellaneous compounds. A number of racemic drugs, including chlormezanon, oxapadol, ketamin, and mianserin, have been resolved on MCTA columns with 96% ethanol as the mobile phase.[29] Poly[N-(S)-acryloylphenylalanine] ethyl ester has been successfully used for optical resolution of chiral sulfur compounds and chlorthalidone with toluene:dioxane (1:1) mobile phase.

Gas Chromatography

As discussed in Chapter 4, cyclodextrin-based phases have been found most useful for gas chromatography of chiral compounds. This has also been found to be the case with preparative separations covered in chapter 9. It is important to recall that a number of compounds can be separated preferably by GC. Such compounds are generally nonaromatic and have a boiling point below 260°C. It is more desirable if these compounds have functional groups such as —OH, —NH$_2$, —COOH, which can be easily derivatized. Some of these commonly used phases (Astec, Whippany, NJ) listed below can be classified into two major categories based on the type of derivatized cyclodextrin used in them.

β-Cyclodextrin-Based Phases

Hydroxypropyl β-cyclodextrin (B-PH). The B-PH phase is especially suitable for saturated analytes with minimum functionality, saturated cyclics, and bicyclics. This series of columns shows less need for inclusion complexation for chiral recognition than the DA columns.

Dialkyl β-cyclodextrin (B-DA). Inclusion complexation or proper fit between the analyte and cyclodextrin cavity is the dominant enantioselectivity mechanism for the DA series of columns. There must be an includable group α or β to the stereogenic center for chiral recognition. This series of columns most effectively separates multiring compounds, one of which is unsaturated (aromatic) α,β to the stereogenic center, such as Prozac, methyl phenidate, and chlorpheniramine. The analysis temperatures are often higher than 150°C. Enantioselectivity has been observed at temperatures greater than 200°C for the Prozac acetyl derivative.

Aromatic alcohols, amino alcohols, aromatic amines, carbohydrates, barbitals, and nicotine analogs have been resolved on Chiraldex B-DA column.

Permethylated β-cyclodextrin (B-PM). The B-PM CSP can be used as a general-purpose column for the separation of acids, alcohols, barbiturates, diols, epoxides, esters, hydrocarbons, ketones, lactones, and terpenes. Some alcohols, diols, and tertiary amines can be better resolved on this phase.

Dimethyl β-cyclodextrin (B-DM). The B-DM is a general-purpose column. Based on the results of some studies, B-DM has been shown to be superior to B-PM and B-PH. It is the column of choice when the elution temperature exceeds 200°C.

γ-Cyclodextrin-Based Phases

Trifluoroacetyl γ-cyclodextrin (G-TA). The G-TA CSP has the highest selectivity for alcohols, diols, and polyols as the free alcohol and as an acyl derivative; amines as acyl derivatives; amino alcohols, amino acids, hydroxy acids; and lactones, furans, and pyrans.

Propionyl γ-cyclodextrin (G-PN). The G-PN column has the highest selectivity for epoxides, aromatic amines (e.g., amphetamine and methylamphetamine), >C6 alcohols, and lactones.

Butyryl γ-cyclodextrin (G-BP). The G-BP column can be used as a general-purpose column, but it is especially useful for amino acids and can be used as a substitute for Chirasil-Val. In general, stationary phases that have an external surface interaction as the dominant chiral recognition mechanism (G-BP, G-PN, and G-TA) provide better chiral efficiency than do inclusion-dominant stationary phases (B-DA, B-DM, B-PH, and B-PM). For inclusion-type separations, the β form is most selective, while the γ form has greater capability for the external adsorption-dominant stationary phases. CSPs from other manufacturers and their applications are given in the Miscellaneous section, under Applications.

Applications

At times, better separations can be obtained by GC. This is exemplified by comparing separations by GC and HPLC (see Figure 10.20) for selegiline, an antiparkinsonism drug sold as Deprenyl, on cyclodextrin-based columns.[32] The retention times for both separations is less than 10 minutes; however, it may be noticed that peaks by GC are both sharper and more symmetrical. Other examples of pharmaceutical compounds are discussed in the following text.

Amino acids. An enantiomer-labeling technique has been designed and used mainly for amino acid analysis by GC. The method is based in part on the fact that a mixture of naturally occurring protein L-amino acids is separable from the corresponding D-enantiomers as N,O,S-PFP isopropyl ester derivative on Chirasil-Val column by temperature programming. The enantiomer-labeling technique com-

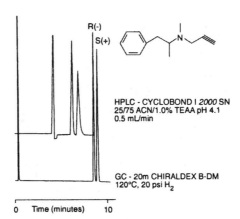

Figure 10.20   GC and HPLC resolution of selegiline. Reproduced with permission from *Chiraldex Columns*, Advanced Separation Technologies, Whippany, NJ. Copyright Advanced Separation Technologies.

pensates for losses during workup and derivatization, but does not account for possible racemization losses of L-amino acids.[33]

An impressive separation by GC of 19 enantiomeric amino acid pairs is shown in Figure 10.21.[34] The degree of racemization can be readily obtained in a separate run under identical conditions with a mixture of pure standards. The advantage of this technique is that it can compensate for the loss of acid-labile amino acids such as tryptophan, cysteine, threonine, and serine during protein analysis.

Natural compounds. CycloSil-B, 30% heptakis (2,3-di-O-methyl-6-O-t-butyldimethylsilyl)-β-cyclodextrin in DB 1701 has recently been introduced by J and W (Folsom, CA). It can resolve a large number of natural compounds, such as the following:

Camphene	Limonene
Camphor	Linalool
α-Ionone	Menthol
Isoborneol	α-Pinene
Isomenthone	

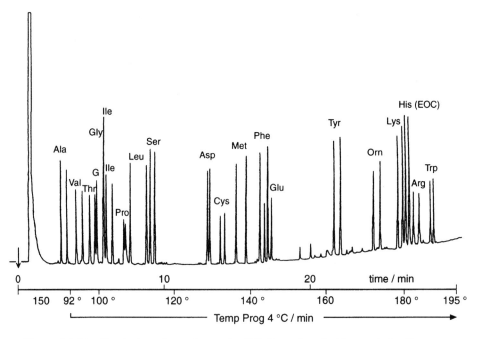

Figure 10.21 Resolution of enantiomers by GC. Reproduced with permission from E. Bayer and K. Frank, in *Modification of Polymers*, C. Carraber Jr. and M. Tsuda, Eds., American Chemical Society, Washington, DC, 1980. Copyright 1980 American Chemical Society.

It can also resolve a number of γ-lactones, alcohols, and diols. Figure 10.22 shows the resolution of a number of components in rosemary oil. Permethylated derivatives of β-cyclodextrin in cyanopropyl-dimethylpolysiloxane liquid stationary phase are commonly used for chiral separations. Limitations in this stationary phase can be circumvented, at times, by using β-cyclodextrins derivatized with alkyl substituents. Various combinations of alkylated β-cyclodextrins into a cyanopropyl-dimethylpolysiloxane liquid stationary phase have been utilized in capillary columns to bring about improvement in separations.[35] Additional separations of some other natural compounds may be found in the Miscellaneous section below.

Pharmaceutical compounds. A number of pharmaceutical compounds such as hexobarbital (*Rs* is 11.3 on Restek's βDEX sm column), mephobarbital (*Rs* is 8.4 on βDEX sm column), and fenfluramine (*Rs* is 2.90 with peak tailing on Restek's βDEX se column) have also been resolved on the indicated columns. Derivatization with TFAA can lead to better resolution without peak tailing for fenfluramine on Restek's βDEX cst column.

Mephenytoin ($R = CH_3$, $R_1 = C_2H_5$, $R_2 = C_6H_5$) can be resolved by GC on a Chirasil-Val column for a pharmacokinetic study.[36] The method allows observation of the enantiomers of mephenytoin, its desmethyl metabolite (derivatized to its 3-propyl homologue), and the internal standard in one chromatogram. It has been found from samples taken at different times that the elimination half-life of the (S)-enantiomer of mephenytoin is less than 3 hours, whereas that for the (R)-form is more than 70 hours.

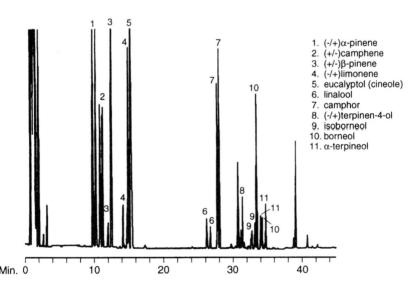

Figure 10.22 Analysis of rosemary oil. Reproduced with permission from *A Guide to the Analysis of Chiral Compounds*, GC Restek Corp., Bellefonte, PA. Copyright Restek.

Miscellaneous compounds. A few examples of a variety of compounds that have been resolved on the commercial cyclodextrin columns from Restek (Bellefonte, PA) are given in Table 10.14. The column that provides the best resolution is also given. Unfortunately the exact chemical structure is proprietary.

Macherey Nagel (Duren, Germany) has introduced fused silica capillary columns based on α-, β-, and γ-cyclodextrins that can provide separations for a large number of compounds.

Methods have been detailed that can be useful for determining the enantiomeric composition of natural flavors and essential oils in plants, herbs, grains, fruits, and beverages.[37] The methods use solid-phase microextraction (SPME) to extract the components of interest, followed by capillary gas chromatography on derivatized cyclodextrin columns for the analysis of flavor components in food products (Table 10.15). Separations of a variety of compounds by GC are shown in Table 10.16 to show broad applicability of this technique.

Supercritical Fluid Chromatography (SFC)

The moderate working temperatures used in SFC offer significant advantages over GC. SFC also offers advantages over HPLC in terms of analysis time because of high self-diffusion coefficient of mobile phases. Furthermore, selection of detectors can include those used in GC. The effect of alcohol concentration on selec-

Table 10.14 Compounds Separated on Restek Columns

Structure	RS	Column
Alcohols		
Linalool	6.0	βDEX sm
Isoborneol	4.0	βDEX sm
Menthol	2.2	βDEX ns
Epoxides		
Styrene oxide	10.2	βDEX se
Esters		
Ethyl-2-methyl butyrate	3.5	βDEX se
Linalylacetate	2.2	βDEX se
Ketones		
Carvone	1.3	βDEX sm
Camphor	4.3	βDEX βDEX sa
Lactones		
γ-Nonalactone	5.9	βDEX cst
δ-Decalactone	2.7	βDEX cst
Terpenes		
α-Pinene	3.5	βDEX sm
Limonene	7.3	βDEX sm

Table 10.15 Cyclodextrin-Based CSPs for GC

CSP	Classes of Compounds
Hexakis-(2,3,6-tri-O-pentyl)-α-cyclodextrin	Carbohydrates, polyols, diols, hydrocarboxylic acid esters, (epoxy)-alcohols, glycerol derivatives, spiroacetals, ketones, and alkyl halides
Hexakis-(2,6-di-O-pentyl-3-O-acetyl)-α-cyclodextrin	Lactones, diols (cyclic carbonates), aminols, aldols (O-TFA), glycerol derivatives (cyclic carbonates)
Heptakis-(2,3,6-tri-O-pentyl)-β-cyclodextrin	Alcohols, cyanhydrins, olefins, hydoxycarboxylic acid esters, alkyl halides
Heptakis-(2,6-di-O-pentyl-3-O-acetyl)-α-cyclodextrin	Amines (TFA), aminols (TFA), trans-cycloalkane-1,2-diols, trans-cycloalkanes-1,3-diols (TFA), β-amino acid esters
Octakis-(2,6-di-O-pentyl-3-O-butyl)-γ-cyclodextrins	α-Amino acids, α- and β-hydroxycarboxyllic acid esters, alcohols (TFA), diols (TFA), ketones, pheromones (cyclic acetals), amines, alkyl halides, lactones
In addition, phases diluted with OV-17 have been found useful	
CSP	Classes of Compounds
Heptakis-(2,3,6-tri-O-methyl)-β-cyclodextrin	Alcohols, diols, hydroxycarboxylic acid esters, olefins, lactones, acetals
Heptakis-(2,6-di-O-methyl-3-O-pentyl)-β-cyclodextrin	Terpenes, dienes, allenes, terpene alcohols, 1, 2- epoxyalkanes, carboxylic acids (esters), hydroxycarboxylic acid esters, pharmaceuticals, pesticides

tivity in SFC is shown in Figure 10.23 for *p*-toluic acid amide derivative of 2-aminoheptane.[38]

A maximum value of α is seen as a function of alcohol concentration for Chiralcel OB column for the tested solute. The modifier regulates the polarity and selectivity of the mobile phase and competes with the solute for the active sites of the stationary phase. These opposing effects produce a maximum for the α value. Furthermore, it may be observed that linear alcohols provide less selectivity than branched alcohols. It is important to recognize that solvation properties of supercritical fluids influence the capacity factor and produce selectivities different from

Table 10.16 Separations of a Variety of Enantiomers by GC

Sample	Column	Temperature (°C)
Amphetamine	Chiraldex G-PN (30 m)	130
Methamphetamine	Chiraldex G-PN (30 m)	130
Pseudoephedrine	Chiraldex G-PN (30 m)	130
Ibuprofen esters	Chiraldex B-PH (10 m)	100
Methyl phenidate	Chiraldex B-DA (30 m)	165
Prozac	Chiraldex B-DA (20 m)	205
Pinene	Chiraldex G-TA (40 m)	34
Camphor	Chiraldex B-TA (40 m)	85
Menthol	Chiraldex G-BP (40 m)	95

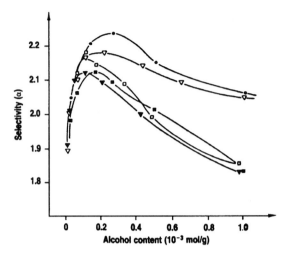

Figure 10.23 Effect of alcohol in SFC. Reproduced with permission from P. Macaudiere, M. Caude, R. Rosset, and A. Tambutte, *J. Chromatogr. Sci.*, **27**, 583 (1989). Copyright 1989 J. Chromatography.

those measured under GC or HPLC using the same column. A comparison of analysis times of oxazepam by HPLC and subcritical fluid chromatography can be seen in Figure 4.8. It can be observed that analysis time is significantly lower. A much lower analysis time is possible with SFC.

A few examples of drugs that can be resolved by SFC are given in Table 10.17.[39–41] Other examples of separations by SFC can be found later in this chapter, chapter 9, and recently published reviews.[42–44]

Capillary Electrophoresis (CE)

Capillary electrophoresis is a useful technique that can achieve rapid, high-resolution separation of water-soluble components present in small-sample volumes. It is useful for chiral separations, and a number of applications have been found. However, the primary limitation is the small sample size used for these separations.

Table 10.17 SFC of Drugs

Drug	Column	Mobile Phase
Ibuprofen[a]	Octanyl-Chirasil-Dex	Carbon dioxide
Mephentoin[b]	DCP cross-linked phase	Carbon dioxide
Warfarin[c]	Chirasil-Dex	Carbon dioxide

a. Z. Juvancz and K. E. Markides, *LC-GC Intl.*, 5, no. 4, 44.

b. D. Johnson, J. Bradshaw, M. Eguchi, B. Rossiter, M. Lee, P. Petersson, and K. Markides, *J. Chromatogr.*, **594**, 283 (1992).

c. V. Schurig, D. Schmalzing, M. Schleimer, and M. Jung, personal communication, 1994.

Method Selection and Selected Applications

Table 10.18 Enantiomers Separations by CE

Enantiomers	Mode	Medium
Amino acids, dansylated	CE	Copper(II)-histidine (or Aspartame)
	MECC	Bile salt micelles
	CGE	β-CD
Binaphthyl enantiomers	MECC	Bile salt micelles
Trimetoquinol, tetrahydropapaveroline, diltiazem	MECC	Bile salt micelles
N-Acylated amino acid esters	MECC	N-Dodecanoyl-L-amino-acidates, Na salt micelles
Phenylthiohydantoin amino acids	MECC	Mixed digitonin-NADS micelles
Terbutaline and propranolol	CD-EC	D-O-M-βCD
Norephedrine, ephedrine, norepinephrine, epinephrine, isoproterenol	CD-CE	D-O-M-βCD
Thiopental, pentobarbital, phenobarbital, barbital, dinaphthyls	Mixed MECC/ CD-CE	NaDS micelles plus cyclodextrins plus other chiral additives

Note: Also review I. E. Valko', H. A. H. Billiet, H. A. L. Corstjens, J. Frank, *LC-GC Int.*, 6, 420, (1993).

Various modes of capillary electrophoresis have been used for resolving a large number of compounds (Table 10.18). The medium used for these separations is also shown in the table. Two other examples are described in some detail here. Propanolol has been resolved in 50 mM phosphate buffer (pH 2.5) containing 4 M urea, 40 mM β-cyclodextrin, and 30% methanol.[45] The effect of increasing concentrations of methanol can be seen in the electropherogram (Figure 10.24). It has

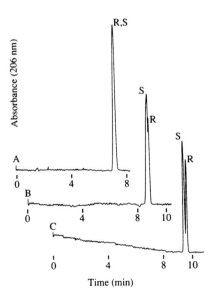

Figure 10.24 Electopherograms of propanolol. Reproduced with permission from S. Fanali, *J. Chromatogr.*, 545, 437 (1991). Copyright 1991 J. Chromatography.

been shown that the addition of 2% to 10% of maltodextrins (mixture of linear α-(1-4)-linked D-glucose polymers) can be useful for chiral resolution of nonsteroidal anti-inflammatory drugs such as flubiprofen, ibuprofen, and naproxen.[46]

Nishi et al.[47] have used bile salts to resolve a variety of chiral molecules, including diltiazem, trimetoquinol, and carboline derivatives. Generally, four bile salts (sodium cholate, sodium deoxycholate, sodium taurocholate, and sodium deoxytaurocholate) are used under neutral-to-alkaline conditions. The enantiomers of diltiazem, its related compounds, and trimetoquinol could be resolved only at pH 7.0 using sodium deoxytaurocholate. The method was used to determine the optical purity of five batches of trimetoquinol down to 1% of the (R) isomer in the presence of (S) isomer (Figure 10.25).

The comparison of separations obtained by CE with other chromatographic techniques is discussed in the next section of this chapter. A number of recent publications dealing with a variety of separations may be useful.[48-54] These publications cover method development strategies, the use of cyclodextrins, and macro-

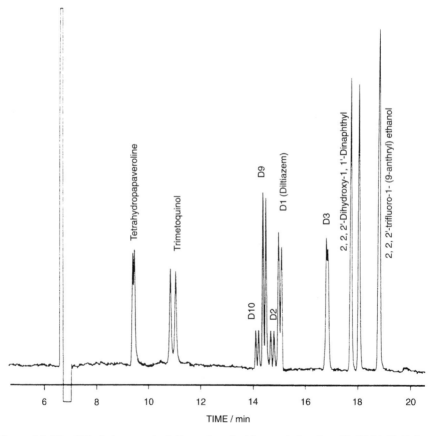

Figure 10.25 CE of trimetoquinol. Reproduced with permission from H. Nishi, T. Fukuyama, and S. Terabe, *J. Chromatogr.*, **515**, 233 (1990). Copyright 1990 J. Chromatography.

cyclic antibiotics in chiral separations. For example, Lin et al.[52] provide a useful list of separations of 63 drugs with seven cyclodextrins by CE.

Comparative Separations

It may be instructive to compare separations obtained by various chromatographic techniques. For example, a comparison of GC, SFC, and CE has been reported for hexobarbital[55] on the same column (1 m × 0.05 mm i.d. fused-silica capillary coated with Chirasil-Dex; the film thickness was 0.25 μm). All three techniques provided good resolution within approximately 30 minutes. It appears that GC provides baseline resolution at 155°C; whereas SFC and CE can be carried out at lower temperatures — 60°C and RT, respectively. The stationary phase can handle mobile phases such as supercritical CO_2 for SFC and an aqueous buffer for CE.

Schurig et al.[56] have evaluated enantioselective separations on capillary Chirasil-Dex columns (85 cm × 0.05 mm i.d. with a film thickness of 0.15 μm) by GC, SFC, HPLC with open tubular coated column and capillary electrochromatography (CEC). As a result of a different diffusion coefficient (D_m) in gases and liquids, analysis time and optimum efficiency in capillary HPLC and CEC are longer by four orders of magnitude, as compared to GC (Figure 10.26). However,

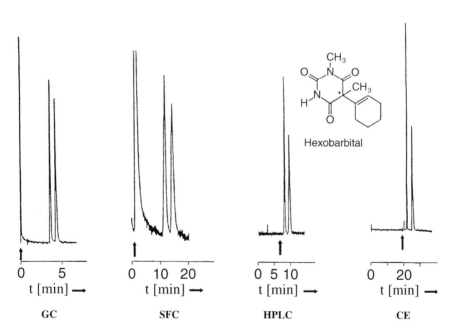

Figure 10.26 Enantiomeric separation of hexobarbital by GC, SFC, HPLC, and CE. Reproduced with permission from V. Schurig, S. Mayer, M. Jung, M. Fluck, H. Jakubetz, A. Glausch, and S. Negura, in *The Impact of Stereochemistry on Drug Development and Use*, H. Y. Aboul-Enein and I. W. Wainer, Eds., Wiley, New York, 1997. Copyright 1997 John Wiley & Sons.

minimum plate height, optimum velocity, and total analysis time depend on the retention factor. The retention factor is large in GC and intermediate in SFC and very small in HPLC and CEC for the first eluted enantiomer.

The following observations can be made on reviewing these separations:

- α for CEC.HPLC > SFC > GC
- R_s for CEC > HPLC > GC > SFC
- N (first peak) for CEC > HPLC = GC > SFC

It should be noted that CEC is the slowest method, due to reduced electro-osmotic flow in the coated capillary.

REFERENCES

1. S. Ahuja, *Chiral Separations: Applications and Technology*. American Chemical Society: Washington, DC, 1997.
2. H. Y. Aboul-Enein and I. W. Wainer, eds., *The Impact of Stereochemistry on Drug Development and Use*. Wiley: New York, 1997.
3. The Innovative Directions in Chiral Separations, *Phenomenex Bulletin*, Torrance, CA.
4. S. Perrin and W. Pirkle, in *Chiral Separations by HPLC*, S. Ahuja, Ed. American Chemical Society Symposium Series #471. Washington, DC, 1991, page 43.
5. *Chiral HPLC Application Guide*, Regis Technologies: Morton Grove, IL.
6. K. W. Street, Jr., *J. Liq. Chromatogr.*, **10**, 655 (1987).
7. *Chiral Separations*, Advanced Separations Technologies: Whippany, NJ.
8. D. W. Armstrong, C. Tang, S. Chen, Y. Zhou, C. Bagwill, J-R. Chen, *Anal. Chem.*, **66**, 1473 (1994).
9. D. W. Armstrong, Y. Liu, and K. H. Ekberg-Ott, *Chirality*, **7**, no. 6 (1995).
10. S. Chen, Y. Liu, D. Armstrong, P. Victory, and B. Martinez-Teipel, *J. Liq. Chromatogr.*, **18**, 1495 (1995).
11. *Chirobiotic Handbook*, Advanced Separation Technologies: Whippany, NJ.
12. C. Dalgliesh, *J. Chem. Soc.*, 137 (1952).
13. *Application Guide for Chiral Column Selection*, Chiral Technologies: Exton, PA.
14. ChromTech Application Note no. 13, 1994.
15. S. Ahuja, presented at the First International Symposium on Separation of Chiral Molecules, Paris, 1988.
16. V. Davankov and A. Kurganov, *Chromatographia*, **13**, 339 (1980).
17. S. Weinstein, M. Engel, and P. Hare, *Analyt. Biochem.*, **121**, 370 (1982).
18. R. Watherby, R. Allan, and G. Johnson, *J. Neuroscience Methods*, **10**, 23 (1984).
19. S. Ahuja, *Chromatographia*, **34**, 411 (1992).
20. I. Wainer, T. Doyle, K. Donn, and J. Powell, *J. Chromatogr.*, **306**, 405 (1984).
21. E. Kusterand and D. Giron, *J. High Res. Chromatogr.*, **9**, 531 (1986).
22. G. Schill, I. Wainer, and S. Barkan, *J. Liq. Chromatogr.*, **9**, 641 (1986).
23. P. Erlandsson, I. Marle, L. Hansson, R. Isaksson, C. Pettersen, and G. Pettersen, *J. Am. Chem. Soc.*, **112**, 4573 (1990).
24. S. Allenmark and S. Andersen, *Chromatographia*, **31**, 429 (1990).
25. Z. Yang, S. Barkan, C. Brunner, J. Weber, T. Doyle, and I. Wainer, *J. Chromatogr.*, **324**, 444 (1985).
26. S. Allenmark, *Chromatographic Enantioseparation*. Ellis Horwood: New York, 1991.
27. W. Pirkle, J. Finn, J. Schreiner, and B. Hamper, *J. Am. Chem. Soc.*, **103**, 3964 (1981).

28. G. Blaschke and J. Maibaum, *J. Pharm. Sci.*, **74**, 438 (1985).
29. G. Blaschke, *J. Liq. Chromatogr.*, **9**, 341 (1986).
30. W. Pirkle and A. Tsipouras, *J. Chromatogr.*, **291**, 291 (1984).
31. S. Allenmark, *J. Liq. Chromatogr.*, **9**, 425 (1986).
32. *Chiraldex Columns*, Advanced Separation Technologies: Whippany, NJ.
33. N. Blair and W. Bonner, *J. Chromatogr.*, **198**, 185 (1980).
34. E. Bayer and K. Frank, in *Modification of Polymers*, C. Carraber Jr. and M. Tsuda, Eds., American Chemical Society: Washington, DC, 1980.
35. *A Guide to the Analysis of Chiral Compounds*, GC. Restek Corp.: Bellefonte, PA.
36. P. Wedlund, B. Sweetman, C. McAllister, R. Branch, and G. Wilkinson, *J. Chromatogr.*, **307**, 121 (1984).
37. V. Mani and C. Woolley, *LC-GC*, **13**, 734 (1995).
38. P. Macaudiere, M. Caude, R. Rosset, and A. Tambutte, *J. Chromatogr. Sci.*, **27**, 583 (1989).
39. Z. Juvancz and K. E. Markides, *LC-GC Intl.*, **5**, no. 4, 44.
40. D. Johnson, J. Bradshaw, M. Eguchi, B. Rossiter, M. Lee, P. Petersson, and K. Markides, *J. Chromatogr.*, **594**, 283 (1992).
41. V. Schurig, D. Schmalzing, M. Schleimer, and M. Jung, personal communication, 1994.
42. K. Anton and C. Berger, *Supercritical Fluid Chromatography with Packed Columns*. Marcel Dekker: New York, 1997.
43. R. E. Majors, *LC-GC*, **15**, 412 (1997).
44. C. Wolff and W. H. Pirkle, *LC-GC*, **15**, 382 (1997).
45. S. Fanali, *J. Chromatogr.*, **545**, 437 (1991).
46. A. D'Hulst and N. Verbeke, *J. Chromatogr.*, **608**, 275 (1992).
47. H. Nishi, T. Fukuyama, and S. Terabe, *J. Chromatogr.*, **515**, 233 (1990).
48. II Introduction to the Theory and Application of Chiral Capillary Electrophoresis, Beckman Instruments: Fullerton, CA. 1993
49. S. Rabel and J. Stobaugh, *Pharm. Res.*, **10**, 171 (1993).
50. K. D. Altria, *Capillary Electrophoresis Guidebook*. Humana: Totowa, NJ, 1995.
51. D. Heiger, R. Majors, and R. Lombardi, *LC-GC*, **15**, 14 (1997).
52. B. Lin, X. Zhu, B. Koppenhoefer, and U. Epperlein, *LC-GC*, **15**, 40 (1997).
53. A. Guttmann, S. Brunet, and N. Cooke, *LC-GC*, **14**, 32 (1996).
54. T. J. Ward, *LC-GC*, **14**, 886 (1996).
55. V. Schurig, *J. Chromatogr.*, **666**, 111 (1994).
56. V. Schurig, S. Mayer, M. Jung, M. Fluck, H. Jakubetz, A. Glausch, and S. Negura, In *The Impact of Stereochemistry on Drug Development and Use*, H. Y. Aboul-Enein and I. W. Wainer, Eds. Wiley: New York, 1997, p. 401.

APPENDIX

Solvents

A list of solvents that may be used for HPLC are given in Table A1. Included in the table is the following useful information:

- UV cutoff
- Refractive index
- Boiling point
- Polarity parameter
- Solvent strength parameter
- Selectivity group
- Water solubility of a given solvent
- Viscosity

Columns for Chiral Separations

Table A2 lists columns that are commercially available for chiral separations. The columns used for preparative separations are provided in Chapter 9. For the latest listing of columns for each group, the reader would be well advised to contact the specific manufacturer. Some of these are listed here.

Advanced Separation Technologies Inc., Whippany, NJ

Chiral Technologies, Exton, PA

Phenomenex, Torrance, CA

Regis Technologies, Morton Grove, IL

J & W Scientific, Folsom, CA

Restek Corp., Bellefonte, PA

Chiroptical Detectors

At least three dedicated HPLC polarimetric detectors are commercially available.

1. SHODEX OR 1 optical rotation detector ($20,000)
   U.S. distributor: JM Science, Inc.
   5820 Main Street, Suite 300
   Buffalo, NY 14221
   tel. (716) 633-3224
   fax (716) 634-1970

2. ACS Chiramonitor Model 2000 (released 1993) $20,000
   U.S. distributor: Polymer Laboratories, Ltd.
   Amherst Fields Research Park
   160 Old Farm Road
   Amherst, MA 01002
   tel. (413) 253-9554
   fax (413) 253-2476

3. Laser diode/fixed wavelength 820 nm
   Polar monitor ($14,000)
   Tungsten halogen source/polychromatic or monochromatic (e.g., 426 nm) 40-μl cell, 220 V AC
   Biichi Laboratoriums-Technik AG
   Postfach
   9230 Flawil
   Switzerland
   41/071 848181
   fax 41/071 835711

4. Vendors of Convertible Chiroptical Spectrometers
   DIP 370 variable wavelength polarimeter with 1 × 50 mm HPLC flow cell ($20,000), laser diode source available

   J 700 series CD spectrometers, scanning/variable wavelength 170–800 nm, NPLC cell 16-μl volume 5-mm path length (>$60,000)
   Jasco, Inc., 8649 Commerce Drive
   Easton, MD 21601
   tel. (900) 333-5272
   fax (410) 822-7526

   Digital polarimeters
   Flow-through sample cells
   Rudolph Instruments, Inc.
   40 Pier Lane
   Fairfield, NJ 07006
   (201) 227-0139

Autopol line of digital polarimeters
Flow-through sample cells
Rudolph Research
One Rudolph Road
P.O. Box 1000
Flanders, NJ 07836
(201) 691-1300
(201) 691-5480

Table A1 Properties of Solvents Used in HPLC

Solvent[a]	Use[b]	UV Cutoff[c]	R.I.[d]	Boiling Point (°C)	Viscosity (cP, 25°)	P'[e]	ε°[f]	Selectivity Group[g]	Water Solubility in Solvent[h]
Isooctane* (2,2,4-trimethylpentane)	LC	197	1.389	99	0.47	0.1	0.01	—	0.011
n-Heptane*	LC	195	1.385	98	0.40	0.2	0.01	—	0.010
n-Hexane	LC	190	1.372	69	0.30	0.1	0.01	—	0.010
n-Pentane	LC	195	1.355	36	0.22	0.0	0.00	—	0.010
Cyclohexane	LC	200	1.423	81	0.90	-0.2	0.04	—	0.012
Cyclopentane	LC	200	1.404	49	0.42	-0.2	0.05	—	0.014
1-Chlorobutane	LC	220	1.400	78	0.42	1.0	0.26	VIa	
Carbon tetrachloride	LC	265	1.457	77	0.90	1.6	0.18	—	0.008
Triethylamine	LC		1.398	89	0.36	1.9	0.54	I	
i-Propyl ether	LC	220	1.365	68	0.38	2.4	0.28	I	0.62
Toluene	LC	285	1.494	110	0.55	2.4	0.29	VII	0.046
p-Xylene	LC	290	1.493	138	0.60	2.5	0.26	VII	
Chlorobenzene	LC		1.521	132	0.75	2.7	0.30	VII	
Ethyl ether	LC	218	1.350	35	0.24	2.8	0.38	I	1.3
Benzene	LC	280	1.498	80	0.60	2.7	0.32	VII	0.058
n-Octanol	LC	205	1.427	195	7.3	3.4	0.5	II	3.9
Methylene chloride	LC	233	1.421	40	0.41	3.1	0.42	V	0.17
i-Pentanol	LC	—	1.405	130	3.5	3.7	0.61	II	9.2
1,2-Dichloroethane	LC	228	1.442	83	0.78	3.5	0.44	V	0.16
t-Butanol	LC	—	1.385	82	3.6	4.1	0.7	II	miscible
n-Butanol	LC	210	1.397	118	2.6	3.9	0.7	II	20.1
n-Propanol	LC	240	1.385	97	1.9	4.0	0.82	II	miscible
Tetrahydrofuran*	LC	212	1.405	66	0.46	4.0	0.57	III	miscible
Propyl amine*	LC	—	1.385	48	0.35	4.2		I	miscible
Ethyl acetate*	LC	256	1.370	77	0.43	4.4	0.58	VIa	9.8
i-Propanol	LC	205	1.384	82	1.9	3.9	0.82	II	miscible
Chloroform*	LC	245	1.443	61	0.53	4.1	0.40	VIII	0.072
Acetophenone	LC	—	1.532	202	1.64	4.8	—	VIa	—
Methylethyl ketone*	LC	329	1.376	80	0.38	4.7	0.51	VIa	23.4
Cyclohexanone*	LC	—	1.450	156	2.0	4.7	—	VIa	—

224

Solvent		Wavelength[c]	Refractive index[d]	BP	% water[h]	Polarity[e]	Strength[f]	Group[g]	Miscibility
Dioxane	LC	215	1.420	101	1.2	4.8	0.56	VIa	miscible
Pyridine	LC	—	1.507	115	0.88	5.3	0.71	III	miscible
Acetone[a]	LC	330	1.356	56	0.30	5.1	0.56	VIa	miscible
Benzyl alcohol	LC	—	1.538	205	5.5	5.7	—	IV	—
Methoxyethanol	LC	210	1.400	125	1.60	5.8	—	III	miscible
Tris(cyanoethoxy) propane	GC	—	—	—	—	6.6	—	VIb	—
Propylene carbonate	LC	—	—	—	—	6.1	—	VIb	—
Ethanol	LC	210	1.359	78	1.08	4.3	0.88	II	miscible
Oxydipropionitrile	GC	—	—	—	—	6.8	—	VIb	—
Acetic acid	LC	—	1.370	118	1.1	6.0	—	IV	miscible
Acetonitrile[a]	LC	190	1.341	82	0.34	5.8	0.65	VIb	miscible
N,N-dimethylacetamide	LC	268	1.436	166	0.78	6.5	—	III	—
Dimethylformamide	LC	268	1.428	153	0.80	6.4	—	III	miscible
Dimethylsulfoxide	LC	268	1.477	189	2.00	7.2	0.75	III	miscible
N-methyl-2-pyrrolidone	LC	285	1.468	202	1.67	6.7	—	III	—
Methanol[a]	LC	205	1.326	65	0.54	5.1	0.95	II	miscible
Nitromethane	—	380	1.380	101	0.61	6.0	0.64	VII	2.1
N-methylformamide	—	—	1.447	182	1.65	6.0	—	III	miscible
Ethylene glycol	—	—	1.431	182	16.5	6.9	1.11	IV	miscible
Formamide	—	—	1.447	210	3.3	9.6	—	IV	miscible
Water	LC	—	1.333	100	0.89	10.2	—	VIII	—

Source: L. R. Snyder and J. J. Kirkland, *Introduction to Modern Liquid Chromatography*, Wiley, NY.

a[*]: indicates preferred LC solvent of low viscosity (<0.5 cP), with convenient boiling point (>45°).

b: LC indicates that solvent can be purchased specifically for use in LC from one of following suppliers: Burdick & Jackson, Baker Chemical, Mallinkrodt Chemical, Fisher Scientific, Water Associates, Manufacturing Chemists, Inc.

GC indicates the solvent is used as a gas chromatography stationary phase and can be purchased from companies selling GC columns and phases (these solvents are used as stationary phases in liquid-liquid LC with mechanically held phase.)

c: Approximate wavelength below which solvent is opaque.

d: Refractive index at 25°C.

e: Solvent polarity parameter.

f: Solvent strength parameter for LSC on alumina.

g: See Snyder's triangle diagram in reference given above.

h: Percent weight of water dissolving in given solvent at 20°C, of interest in LSC.

Table A2 Commercial Chiral Columns for HPLC*

Name	Chemical Name	Type	Particle Size	Source
Apex prepsil L-valylphenylurea	(S)-Valyl-phenylurea	Brush	8	Jones
Bakerbond chiral covalent DNBLeu	(S)-DNB-Leucine (covalent)	Brush	5	Baker
Bakerbond chiral covalent DNBPG	(R)-DNB-Phenylglycine (covalent)	Brush	3,5,10	Baker
Bakerbond chiral ionic DNBPG	(R)-DNB-Phenylglycine (ionic)	Brush	5	Baker
Cellulose CEL-AC-40 XF	Cellulose-triacetate	Cellulose	7	Macherey-Nagel
Chiral-AGP	α₁-Acid-glycoprotein	Protein	5	ChromTech
Chiralcel CA-1	Cellulose-triacetate	Cellulose	10	Daicel
Chiralcel OA	Cellulose-triacetate	Cellulose	10	Daicel
Chiralcel OB	Cellulose-tribenzoate	Cellulose	10	Daicel
Chiralcel OC	Cellulose-trisphenylcarbamate	Cellulose	10	Daicel
Chiralcel OO	Cellulose-tris-3,5-dimethylphenylcarbamate	Cellulose	10	Daicel
Chiralcel OF	Cellulose-tris-4-chlorphenylcarbamate	Cellulose	10	Daicel
Chiralcel OG	Cellulose-tris-4-toluylcarbamate	Cellulose	10	Daicel
Chiralcel OJ	Cellulose-tris-4-toluylate	Cellulose	10	Daicel
Chiralcel OK	Cellulose-tricinnamate	Cellulose	10	Daicel
Chiralcel WE	N-(2-Hydroxy-1,2-diphenylethyl)glycine-copper	Ligand exchange	10	Daicel
Chiralpak OT(+)	Poly(triphenylmethyl-methacrylate)	Cellulose	10	Daicel
Chiralpak OP(+)	Poly(2-pyridyl-diphenylmethyl-methacrylate)	Cellulose	10	Daicel
Chiralpak WH	Proline copper	Ligand exchange	10	Daicel
Chiralpak UM	Amino acid copper	Ligand exchange	10	Daicel
Chiral D-DL=Daltosil 100	(R)-DNB-Leucine (covalent)	Brush	4	Serva
Chiral L-DL=Daltosil 100	(S)-DNB-Leucine (covalent)	Brush	4	Serva
Chiral D-DPG=SilOO	(R)-DNB-Phenylglycine (covalent)	Brush	3,5	Serva
Chiral L-DPG=Daltosil 100	(S)-DNB-Phenylglycine (covalent)	Brush	4	Serva
Chiral hypra-Cu=Daltosil 100	Hydroxyproline copper	Ligand exchange	4	Serva
Chiral proCu=SilOO	Proline copper	Ligand exchange	5	Serva
Chiral valCu=SilOO	Valine copper	Ligand exchange	5	Serva
Chiralprotein 1	Beef serum albumin	Protein		SFCC
Chiral protein 2	Human serum albumin	Protein		SFCC
ChiraSpher	Poly-n-acryloyl-(S)-phenylalaninethylester	Cellulose	5	Merck
ChiRSil I	(R)-ONB-phenylglycine (ionic)	Brush	5,10	RSL
Chi-RoSil	(R)-DNB-phenylglycine (ionic)	Brush	5	RSL

Covalent L-leucine	(S)-DNB-leucine (covalent)	Brush	5	Regis, Alltech
Covalent D-naphthylalanine	(R)-Naphthylalanine	Brush	5	Regis, Alltech
Covalent L-naphthylalanine	(S)-Naphthylalanine	Brush	5	Regis, Alltech
Covalent D,L-naphthylalanine	(R,S)-Naphthylalanine	Brush	5	Regis, Alltech
Covalent D-phenyl glycine	(R)-DMB-phenylglycine (covalent)	Brush	5, 10	Regis, Alltech
Covalent L-phenyl glycine	(S)-DMB-phenylglycine (covalent)	Brush	5, 10	Regis, Alltech
Covalent D,L-phenyl glycine	(R,S)-DNB-phenylglycine (covalent)	Brush	5	Regis, Alltech
Crownpak CB	Crown ether	Cavity	10	Daicel
Cyclobond I	β-Cyclodextrin	Cavity	5	Astec
Cyclobond II	γ-Cyclodextrin	Cavity	5	Astec
Cyclobond III	α-Cyclodextrin	Cavity	5	Astec
Cyclobond I-acetylated	Acetylated β-cyclodextrin	Cavity	5	Astec
Cyclobond III-acetylated	Acetylated-cyclodextrin	Cavity	5	Astec
α-Cyclodextrin = Daltosil 100	α-Cyclodextrin	Cavity	4	Serva
β-Cyclodextrin = Daltosil 100	β-Cyclodextrin	Cavity	4	Serva
EnantioPac	α_1-Acid-glycoprotein	Protein	10	Pharmacia
ES D-DMB-LEU	(R)-DNB-Leucine (covalent)	Brush	5	ES
ES L-DNB-LEU	(S)-DNB-Leucine (covalent)	Brush	5	ES
ES D-DNB-PHGLY	(R)-DNB-Phenylglycine (covalent)	Brush	5	ES
ES L-DNB-PHGLY	(S)-DNB-Phenylglycine (covalent)	Brush	5	ES
ES R-PU	(R)-Phenylethylurea	Brush	5	ES
ES S-PU	(S)-Phenylethylurea	Brush	5	ES
Grom-chiral-(R)-DNBPG-C	(R)-DNB-phenylglycine (covalent)	Brush	5	Grom
Grom-chiral-(R)-DNBPG-I	(R)-DNB-phenylglycine (ionic)	Brush	5	Grom
Grom-chiral-(S)-DNBPG-C	(S)-DNB-phenylglycine (covalent)	Brush	5	Grom
Grom-chiral-(S)-DNBPG-I	(S)-DNB-phenylglycine (ionic)	Brush	5	Grom
Grom-chiral-(R)-DNBL-C	(R)-DNB-leucine (covalent)	Brush	5	Grom
Grom-chiral-(R)-DNBL-I	(R)-DMB-leucine (ionic)	Brush	5	Grom
Grom-chiral-(S)-DNBL-C	(S)-DNB-leucine (covalent)	Brush	5	Grom
Grom-chiral-(S)-DNBL-I	(S)-DNB-leucine (ionic)	Brush	5	Grom
Grom-chiral-beta-CD	β-Cyclodextrin	Cavity	5	Grom
Grom-chiral-HP	Hydroxyproline copper	Ligand exchange	5	Grom
Grom-chiral-P	Prolinamide	Ligand exchange	5	Grom

(continued)

Table A2 (Continued)

Name	Chemical Name	Type	Particle Size	Source
Grom-chiral-PC	Proline copper	Ligand exchange	5	Grom
Grom-chiral-PC	Proline copper	Ligand exchange	5	Grom
Grom-chiral-U	(R)-N-α-Phenylethylurea	Brush	5	Grom
Grom-chiral-UC	Valin copper	Ligand exchange	5	Grom
Ionic L-leucine	(S)-DMB-leucine (ionic)	Brush	5	Regis, Alltech
Ionic D-phenyl glycine	(R)-DMB-phenylglycine (ionic)	Brush	5	Regis, Alltech
MCI gel CRSIOW	C_{18}-Silica gel with N,N-dioctyl-(S)-alanine	Ligand exchange		Mitsubishi
Nucleosil chiral-1	Hydroxyproline copper	Ligand exchange	5	Macherey-Nagel
Nucleosil chiral-2	—	Brush	5	Macherey-Nagel
Optimer P1	Aromatic amide	Brush	4	Toyo Soda
Optimer L1	Aliphatic amino acid copper	Ligand exchange	5	Toyo Soda
Optimer L2	Aromatic amino acid copper	Ligand exchange	5	Toyo Soda
Resolvosil	Beef serum albumin	Protein	7	Macherey-Nagel
Spherisorb chiral 1	(R)-N-α-Phenylethylurea	Brush	5	PhaseSep
Spherisorb chiral 2	(R)-Naphthylethylurea	Brush	5	PhaseSep
Sumichiral OA-1000, OA	α-Naphthylethylamide	Brush	5,10	Sumika
Sumichiral OA-2000	(R)-DMB-Phenylglycine (ionic)	Brush	5,10	Sumika
Sumichiral OA-2000A	(R)-DNB-Phenylglycine (covalent)	Brush	5,10	Sumika
Sumichiral OA-2100	Chlorphenyl-isovaleroyl-phenylglycine	Brush	5,10	Sumika
Sumichiral OA-2200	Chrysanthemoyl-phenylglycine	Brush	5,10	Sumika
Sumichiral OA-3000	tert, Butylaminocarbonyl-valine	Brush	5,10	Sumika
Sumichiral OA-4000	(S),(S)-α-Naphthylethyl-aminocarbonyl-valine	Brush	5,10	Sumika
Sumichiral OA-4100	(R),(R)-α-Naphthylethyl-aminocarbonyl-valine	Brush	5,10	Sumika
Supelcosil LC-(R)-naphthylurea	(R)-Naphthylethylurea	Brush	5	Supelco
Supelcosil LC-(R)-urea	(R)-Phenylethylurea	Brush	5	Supelco
Triacetylcellulose	Cellulose-triacetate	Cellulose	10	Merck
Trichsep-100	Cellulase	Protein	10	Sonsep
TSKgel Enantio Li	Aliphatic amino acid copper	Ligand exchange	5	TosoHaas
TSKgel Enatio L2	Aromatic amino acid copper	Ligaiid exchange	5	Tosollaas
TSKgel Enantio PI	(S)-Aromatic amide	Brush	4	TosoHaas
Ultron OVM	Ovomucoid	Protein		Shinwa
VMC-Pak K	Polymer with (R)-naphthylethylamine	Brush	5	YMC

*Source: *Chiral Separations by Liquid Chromatography*, S. Ahuja, Ed.., ACS Symposium Series #471, American Chemical Society: Washington, DC, 1991, pp. 8–13.

INDEX

Abscisic acid, semipreparative separations of, 148–149
Absolute configuration
 biological activity and, 53–55
 determination of, 52–53, 95
Acetonitrile, as solvent, 27, 100, 139, 182–183, 193
Achirality
 in chiral separations, 8–9, 123, 131, 137
 derivatization applications of, 174–175
 superimposibility and, 38–39
α1-Acid glycoprotein column (AGP), 123–124, 192–194
Acid/base concentration. See pH
Acids
 separation methods for, 164, 167
 as stationary phase, 150–152
 See also specific acid
Additives, for chiral separation
 with capillary electrophoresis, 78
 examples of, 8–9, 61–62, 131
 in mobile-phase, 75, 132–135
 in stationary phase, 59–60
Adenylate cyclase, 20
Adrenaline, 19
Adrenoreceptors
 α-adrenoreceptors: one chiral center compounds and, 18–19; two chiral center compounds and, 19–20
 β-adrenoreceptors: blockers of, 55, 202 (see also specific drug); one chiral center compounds and, 17–19; stimulants of,

 absolute configuration of, 53–55; two chiral center compounds and, 19–20
 Easson-Stedman hypothesis for, 19
 eudismic ratio comparisons, 17–20
Adsorption chromatography
 charge-transfer based, 106
 principles of, 9–10, 111, 129
Alanine
 resolution of, 60, 182
 as stationary phase, 109
Albumin binding
 in chromatography, 101
 pharmacokinetic principles of, 25
Alcohols
 chromatography methods for, 64–65
 derivatization of, 64–65, 71–72
 effects on supercritical fluid chromatography, 212–214
 high pressure liquid chromatography and, 199–201
 resolution of, 106, 211
 as solvents, 137–139
 See also specific type
Aldehyde groups, from glycoproteins, 123
Alkaloids, high pressure liquid chromatography and, 199, 201
Alkenes, chromatography methods for, 42, 71
Allenes, molecular structure of, 42, 44
Alprenolol, resolution of, 156, 200, 202
Amfepramone, 29
Amides, high pressure liquid chromatography and, 200–201

229

Amines
  chromatography methods for, 64–65, 171
  derivatization of, 64–65, 102, 113
  high pressure liquid chromatography and, 200, 202
  protein column separations for, 123, 196
  resolution of, 106
Amino acid derivatives
  brush-type, 106–109, 136, 171
  for chiral stationary phases, 102–104, 109
  chromatography methods for, 60–69, 74, 106–108, 113
  racemates of, 60
  reagents for, 60–63, 65, 70
  resolution of, 203
Amino acid esters
  chromatography methods for, 64–65
  derivatization of, 64–65, 68, 137
Amino acids
  chromatography of: methods for, 64–65, 74, 112–114, 129; parameters for, 203–204
  enantiomer-labeling analysis technique, 209–210
  high pressure liquid chromatography and, 200, 203
  resolution of, 106, 129, 132–134, 199–200, 209–210
  as stationary phase, 106–107, 123, 151
  turkey ovomucoid sequence of, 123–124
Amino alcohols
  chromatography methods for, 64–65
  derivatization of, 64–65, 68, 70
  ligand exchange and, 133
  resolution of, 106
Ammonia, in chromatography, 76, 121, 138, 182, 196, 198
Amosulalol, 18–19
Amphetamine, resolution of, 202
Amylopectin, as stationary phase, 96
Amylose, as stationary phase, 96, 99, 151, 192
Analysis time
  of comparative separations, 216–217
  influencing factors, 94, 101, 214
Analytes
  in chiral stationary phases, 135–136, 138
  in enantioselective interactions, 112–114, 120, 123
  high pressure liquid chromatography and, 199–207
Analytical separations, in elution chromatography, 147–148
Anesthetics, resolution of, 162, 202
Angiotensin-converting enzyme inhibitors, 16
Anisotropic radiation, 85
Anthryl carbinol, 106
Antibiotics. *See* Macrocyclic antibiotics

Antibodies, in chromatography, 101
Arago, François, 39
Aromatic rings, in chromatography, 112, 115, 123, 129, 172, 179
Arylalkylamines, as stationary phase, 106, 114
Ascorbic acid, 16
D-Aspartic acid, 29
Assays, pharmacokinetic, guidelines for, 31–32, 35
Association constants, in chromatographic enantioselectivity, 117–120
Asymmetry, molecular
  determination of, 168–170
  stereoisomerism from, 3, 41–43, 45–46
  topological, 41
Atenolol, resolution of, 194, 202
Atropine, resolution of, 201
Atropisomerism, molecular principles of, 42–46
Attractive interactions, in chiral separations, 114–115, 138–139
Automation, of preparative separations, 148–149
Axial chirality, defined, 41, 45
Aziridines, chromatography methods for, 71

Bacillus macerans, 99
Baclofen, resolution of, 195, 198
Barbiturates
  chromatography methods for, 66, 68, 71, 74, 211
  high pressure liquid chromatography of, 134–135, 200, 203–205
  preparative separations of, 155
  See also specific drug
Bases, as stationary phase, 150–152
Batch-to-batch consistency, in pharmaceuticals, 30–31
B-DA (dialkyl β-cyclodextrin phase), 208
B-DM (dimethyl β-cyclodextrin phase), 208
Bendroflumethiazide, preparative separations of, 156, 158
Benzene, and chiral separations, 97, 120–122
Benzodiazepinones
  high pressure liquid chromatography and, 200, 206–207
  preparative separations of, 157
Benzothiadiazines, high pressure liquid chromatography and, 200, 205–206
Benzyl alcohol, absolute configuration of, 53
Beverages, enantiomeric composition of, 212–213
Biaryls, substituted, 43, 45
Bile salts, in capillary electrophoresis, 214
Bioavailability, of drugs, demonstration guidelines, 35

Biological activity
  absolute configuration and, 53–55
  chiral molecules and, 3–6, 17
Biological testing, of enantiomers, 146
Biopolymeric phases. *See* Protein columns
Biot, Jean-Baptiste, 39
Birefringence, defined, 85
Bovine serum albumin (BSA), 123
B-PH (hydroxypropyl β-cyclodextrin phase), 208
B-PM (permethylated β-cyclodextrin phase), 208
Bridging studies, for pharmaceutical development, 33
Bronchodilators, absolute configuration of, 53–55
Brush-type columns (Type 1)
  amide-type, 172, 175
  derivatization for, 174–175
  high pressure liquid chromatography and, 171–177
  stationary phase principles of, 106–109, 136
  urea-type, 172–175
Buffers
  effect on cyclodextrin separations, 181–182, 187
  effect on protein separations, 193–197
  in high pressure liquid chromatography, 84, 100, 138, 182
Bupivacaine, resolution of, 201
α-Burke columns, 174–176
Buthiazide, preparative separations of, 156
Butyryl γ-cyclodextrin phase (G-BP), 209

Caffeine, 16
Cahn-Ingold-Prelog Convention, 12, 41, 47
Camphene, resolution of, 210
Camphor, resolution of, 210
Capacity factor
  and resolution, 93–95, 100, 107
  temperature effect on, 118, 120
Capillary columns
  in gas chromatography, 66, 68–72
  resolution potential of, 119
Capillary electrophoresis (CE)
  advantages of, 77–79, 214
  free solution mode, 77–78
  high pressure liquid chromatography versus, 78
  micellar electrokinetic mode, 77–78
  overview of, 6–8, 58
  pharmaceutical applications of, 215–218
  selected applications of, 212, 214–216
  separation process of, 77–78
Carbamates
  in chromatography, 98–99, 183–184
  as stationary phase, 188, 190–191

Carbohydrates
  derivatization of, 46, 65, 68
  as stationary phase, 96–100
Carbon atom, asymmetric
  in bronchodilators, 53–55
  defined, 3, 12, 46
  determination of, 169–170
  stereoisomerism of, 40, 42–43, 46
Carbon dioxide, in chromatography, 76
Carboxylic acids
  chromatography methods for, 65, 74, 112, 129, 171
  derivatization of, 64–65, 68, 82, 137
  protein column separations for, 123
  resolution of, 116, 133–134
Carboxylic compounds, high pressure liquid chromatography and, 200, 205
Carvedilol, 19
Cellobiohydralase-I (CBH-I), 123
Cellulose
  carbamate, 188, 190–191
  enantioselectivity of, 121–122, 129
  esters, 188, 190–191
  stationary phase principles of, 96–100, 112, 136, 138, 150–151
Cellulose derivatives, for chiral separations, 9–10, 136
Cellulose plates, for chromatography, 60–62
Cellulose triacetate, microcrystalline (CTA-1), for chiral separations, 151–153
Central chirality, defined, 45
Charge-transfer based adsorption, for resolution, 106
Charge-transfer interactions, in chromatographic enantioselectivity, 115–116
Chelates, for chiral stationary phases, 102, 196–199
Chemical separation, for chiral separations, 6, 49
Chiradex columns, 71
Chiral cavities, 130, 138–139
Chiral center
  defined, 3, 12, 40
  determination of, 169–170
  pharmaceutical consideration of, 27–28
  stereoisomerism from, 40–43
Chiral compound plates, for chromatography, 60–63
Chiral compounds
  nomenclature for, 12, 46–48
  overview of, 4–6
  planar, 41, 46, 153, 162
  three-point interaction theory of, 96, 112, 126
Chiral heteroatoms, preparative separations of, 153, 161

Index

Chiral metal complexes, in gas chromatography, 42, 71–74, 102, 122, 133
Chiral molecules
  atropisomerism, 42–46
  biological activity and, 3–6
  classification of, 45–46
  compounds of. *See* Chiral compounds
  configuration of, 12, 40–46
  conformation of, 12, 40–46
  defined, 3, 12
  overall structure, 42, 44
  single asymmetric atom, 42–43
  stereoisomerism principles of, 40–41, 45–46, 111
  twisted conformation, 45–46
  See also Enantiomers
Chiral recognition models, 112–113, 129, 138–139
Chiral separations
  adsorption based, 9–10, 106, 111, 129
  advantages of, 92
  attractive interactions in, 114–115, 138–139
  chromatographic (Chromatography)
  comparative, selected applications of, 216–218
  desirable features of, 92–96, 110–111
  examples of, 71. (*see also specific compound*)
  ion exchange-based, 102, 105, 111
  methods for, 6, 49–52, 58, 90, 128 (*see also specific method*)
  overview of, 6–8, 128–130
  partition-based, 111
  plates for, 60, 62–64, 93–94, 105–106
  preparative, 10–11, 92
  repulsive interactions in, 114–115
  temperature factor, 63–64, 68–73, 117–120
Chiral stationary phases (CSP)
  capacity determination of, 148–149
  in chiral separation, 49–51, 58: gas chromatography and, 65–72; high pressure liquid chromatography and, 75–76, 135, 170–207; method development, 135–140; thin-layer chromatography and, 59–64
  commercially available, 71, 98–99, 109, 135
  comparisons of, 108–109
  desirable features of, 92–96
  overview of, 8–10
  for preparative separations, 148–150
  rational designs of (*see* Rational stationary phase)
  selection procedure for, 135–139, 144, 170–171: inclusion analyses, 136–140; normal-phase analyses, 135–136; reversed-phase analyses, 136, 151

  solvents for, 98–99, 101, 104–105, 107
  three-point rule of, 96, 112, 126, 129
  types of, 96–108, 122–125, 135–136, 150: modern designs, 9–10; Type 1 (*see* Brush-type columns); Type 2 (*see* Polysaccharide columns); Type 3 (*see* Inclusion chiral stationary phase); Type 4 (*see* Ligand exchange columns); Type 5 (*see* Protein columns)
  See also Columns
Chirality
  axial, 41, 45
  central, 45
  molecular structure and (*see* Chiral molecules)
  overview of, 3–5
  pharmaceutical applications of, 4–6
  planar, 41, 46
  stereoisomerism and, 40–41
  superimposibility of, 3, 38–39
  torsional, 41
Chirasil-Val columns, 66, 68
Chirobiotic columns
  applications of, 187, 189, 203–204
  comparability of separations on, 187–189
  principles of, 121, 137, 184–187
Chiroptical detectors
  advantages of, 86–87, 90
  circular dichroism, 50, 52, 85, 88–90, 117
  commercially available, 227–228
  optical rotatory dispersion, 50, 52, 85–91
  overview of, 85–86
  sample detectability, 87–88
  tutorial on, 90–91
Chiroptical methods, for resolution, 50–52
Chirule, 141
Chlormezanone, preparative separations of, 154
Chlorthalidone (CTD)
  preparative separations of, 155
  resolution of, 27, 29
Chromatography, for chiral separations
  adsorption, 9–10, 106, 111, 129
  direct methods, 7–8, 130–131
  displacement, 146–147
  elution, 95–96, 147–167
  enantio-enriched selector molecules for, 49–50
  enantioselective interactions and, 111–120, 129–130, 196–197
  frontal, 146
  gas (*see* Gas chromatography)
  high pressure liquid (*See* High pressure liquid chromatography)
  historical account of, 7–9, 128, 131
  inclusion and, 97, 99, 120–122, 130, 134–135

indirect methods, 7–8, 130–131
ion pairing, 102, 105, 111, 122–123, 133–134, 151
large-scale preparative liquid systems, 10–11, 79, 130–131, 163
liquid (*see* Liquid chromatography)
low pressure liquid, 103, 123
method selection tutorial, 79–80 (*see also* Method selection)
methods overview, 6–8, 49–52, 58, 130–131
micellar electrokinetic, 77–78
micellar electrokinetic capillary, 78
molecular modeling and, 125
paper technique (*see* Cellulose)
pharmaceutical applications of, 59–60, 66
process defined, 111
protein column separations, 122–125
relative resolution power comparisons, 94–95, 216–218
reverse-phase (*see* Reverse-phase liquid chromatography)
selected applications of, 168–218
simulated moving-bed, 152, 163, 166
subcritical fluid, 76–77
supercritical fluid (*see* Supercritical fluid chromatography)
theoretical rationale, 111–127
thin-layer (*see* Thin-layer chromatography)
transition metal complexes and, 42, 71–74, 102, 122, 133
tutorial for, 110, 126–127
Chromophores, molecular, detection methods and, 85, 87, 89, 102
Circular dichroism (CD)
as chiroptical technique, 85, 88–90
polarimetry and, 50, 52
vibrational, 85, 117
Circularly-polarized light (CPE), as chiroptical technique, 85
cis ligand, 41, 47
Clearance, of drugs. *See* Pharmacokinetics
Coenzyme A thioester, 24
Columns, in chiral separation
chirobiotic, 121, 137, 184–189, 203–204
in chromatography, 122–125
classification of, 135–136
commercially available, 70–71, 98–99, 221, 224–226
desirable features of, 9–10, 27, 92–95
in gas chromatography, 66, 68–72
in high pressure liquid chromatography, 170–207
load capacity of, 153
for preparative separations, 148–151, 153
protein-based, 122–125
Restek, 211–212

selection procedure for, 135–140, 144, 168–171
temperature effects on, 118–120
See also Chiral stationary phases
Commercial plates, for chromatography, 60, 63–64
Complementary functionality, 112
Complexation gas chromatography, 71–74
Computer models
for chromatography, 125
of molecular docking, 140
Conductivity detection methods, 82–84
Configuration
absolute, 52–55
defined, 12, 41
stability issues: molecular configuration and, 43, 45; in pharmaceuticals, 27–29
stereoisomer principles of, 41–46
Conformation
defined, 12, 41
displacement of, 117, 119
in gas chromatography, 68
stereoisomer principles of, 41–46, 129
twisted, example of, 45–46
Conventions. *See* Nomenclature systems
Coordination complexes, metal, 42, 71–74, 122
Coordination sphere, 122
Copper, in chromatography, 102, 196–199
Cotton effect, in optical rotation detection, 88–90
Counterions, in chromatography, 134
Cross-linking agents, in polymer synthesis, 103, 105, 122–123
Crown ether columns
enantioselectivity of, 121–122
high pressure liquid chromatography and, 198–199
stationary phase principles of, 104–105, 109, 130
Crystallization
for chiral separations, 6, 39–40, 49, 128
X-ray methods, 53, 115–115
Cyclodextrin derivatives
bonded, 183–187
chromatography applications of, 30, 68–71, 177, 183–185, 211
Cyclodextrins (CD)
α-cyclodextrin: chromatography applications of, 99, 134, 164, 179; structure of, 176–178
β-cyclodextrin: chromatography applications of, 59–61, 134–135, 177–179, 208–212; dialkyl phase (B-DA), 208; dimethyl phase (B-DM), 208; hydroxypropyl phase (B-PH), 208; permethylated phase (B-PM), 208; in preparative

Cyclodextrins (CD) (*continued*)
  separations, 151, 162, 164; sources of, 120–121; structure of, 176–178; in thin-layer chromatography, 60, 62–63
  γ-cyclodextrin: butyryl phase (G-BP), 209; chromatography applications of, 99, 134, 164, 179, 208–212; propionyl phase (G-PN), 209; structure of, 176–178; trifluoroacetyl phase (G-TA), 208
  in gas chromatography, 208–213
  high pressure liquid chromatography and, 176–187: normal-phase separations, 182–184, 187; reversed-phase separations, 178–182, 187
  inclusion complexes with, 97, 99, 120–122, 130, 134–135, 178–179, 182
  macrocyclic antibiotic separations on, 184, 186–187
  pharmaceutical separations on, 184, 186–187, 211–213
  physical properties of, 177–179
  Restek columns of, 211–212
  sources of, 120–121, 176
  as stationary phase: principles of, 99–101, 109, 130, 136, 139, 150–151, 162, 164; separation influences, 181–182
  in thin-layer chromatography, 59–61, 99
Cytochrome P-450 enzymes, 26

D-conventions, defined, 12, 46
d rotation. *See* Dextrorotation
DBCA (5-Dimethyl-sulfamoyl-6,7-dichloro-2,3dihydrobenzofuran-2-carboxylic acid), 25–26
Derivatization methods
  in gas chromatography, 65–72, 131
  in high pressure liquid chromatography, 74–76, 131–132
  with minimal manipulations, 111
Derivatization reagents, for diastereoisomers, 132
  in gas chromatography, 64–65
  in thin-layer chromatography, 59–64
Detection methods
  chiral compound specific, 85–91
  circular dichroism, 85, 88–90
  development of (*see* Method development)
  enantiomeric composition factors, 81–82
  for gas chromatography, 82
  for high pressure liquid chromatography, 83–84
  miscellaneous nonspecific, 83–84
  optical rotation, 85–88
  polarimetric, 49–52, 85–88
  selection of (*see* Method selection)
  for thin-layer chromatography, 81–82
  tutorial for, 90–91

Detectors
  chiroptical, 85–91
  commercially available, 83–84
  ideal characteristics of, 83
  typical responses of, 84
Deuterium labeling technique, pharmacokinetic applications of, 23–24
Dextromethorphan, 5, 14
Dextrorotation (d), defined, 3–4, 12, 85, 128
Dialkyl β-cyclodextrin phase (B-DA), 208
Diastereoisomers
  bonding mechanisms, 114–117
  in chiral separations, 6–9, 41, 68, 114
  defined, 3, 12, 40
Diastereomeric complexes, x-ray crystallography of, 115–117
Diastereomeric derivatives, high pressure liquid chromatography of, 59–65, 74–75, 132
Diastereomers. *See* Diastereoisomers
Diazepam, resolution of, 206–207
Dichloromethane, as solvent, 95
Diethylpropion. *See* Amfepramone
Diffusion coefficient, in chromatography, 216
Diltiazem, resolution of, 215
Dimethyl β-cyclodextrin phase (B-DM), 208
5-Dimethyl-sulfamoyl-6,7-dichloro-2,3dihydrobenzofuran-2-carboxylic acid (DBCA), 25–26
Diols, resolution of, 211
Dioxane, as solvent, 27
Dipeptide phase, for gas chromatography, 66
Dipole-dipole interactions, in chiral separation, 114–115, 121, 129–130
Disopyramide, preparative separations of, 158
Displacement chromatography, 146–147
Distillation, for chiral separations, 49
Distomer
  defined, 12
  pharmaceutical applications of, 17–19
Diuretics, high pressure liquid chromatography and, 200, 205–206
DNB-phenylglycine, as stationary phase, 106–107, 109, 172–173
Dobutamine, 18
L-Dopa, 16
Drug interactions, study approaches for, 26–29
Dynamic stereochemistry, 40

Easson-Stedman hypothesis, for adrenoreceptors, 19
Efficiency, in chiral stationary phases, 93–94, 105, 107, 120
Electrochemical detection method, 83–84
Electrolytes, for capillary electrophoresis, 78

Electron-releasing substituents, in chromatography, 115–116, 121
Electrostatic forces, in chiral separations, 114, 129–130
Elution chromatography
 applications of, 153–167
 automation processes, 148–149
 chiral stationary phases in, 148–151
 defined, 146
 displacement with, 146–147
 inversion process in, 95–96, 99, 101, 107, 152
 method development, 152
 resolution strategy, 148, 151–152
 sample solubility, 148
 scale-up of analytical separations, 147–148
 solvent recovery in, 148–149
Elution factors, in detection methods, 81–83, 216
Elution order
 inversion principles of, 95–96, 99, 101, 107, 152, 174
 prediction techniques, 125
Enantiomeric enrichment
 for chiral separations, 6, 104, 106
 selector molecules for, 49–50
Enantiomeric purity
 biological consequences of, 20–22
 chromatographic determinations of, 52, 114
 importance of, 4–6
 in pharmaceuticals: range of, 15, 20–22, 28–29; rules for, 30–31, 35
 in preparative separations, 10–11, 106
Enantiomeric resolution
 using chiral mobile-phase additives, 132–135
 using chiral stationary phases, 135–140
Enantiomeric stability
 configuration factors of, 12, 41–46
 in pharmaceuticals, 27–31
Enantiomer-labeling technique, for amino acid analysis, 209–210
Enantiomers
 biological activities of, 3–6, 17
 composition determination methods, 81–82
 defined, 3, 38, 40 12
 drug applications of (*see* Pharmaceuticals)
 optical rotation properties of, 3–4, 40
 pathways to, 146, 170–171
 preparative separations of, 10–11, 145–146
 single (*see* Single enantiomers)
 three-point rule of, 96, 112, 126
 See also Chiral molecules

Enantioselectivity, chromatographic
 charge-transfer interactions in, 115–116
 in chiral separations, 111–114, 129–130
 hydrogen bonding in, 115–117, 196
 overview of, 114–115
 pH effects on, 196–198
 temperature effects on, 63, 68, 117–120, 129
Energy, as separation factor
 in molecular modeling, 125
 in stereoisomers, 41, 68, 75
 temperature effects on, 117–120
Enrichment
 enantiomeric, 6, 49–50, 104, 106
 stereoisomeric, 28–29
Enzymatic resolution, for chiral separations, 6
Enzyme catalysis, of enantiomers, 10–11, 49
Epimers, applications of, 12, 28
Epinephrine, 54
Epoxide hydrolase, cytosolic, 26
Epoxides, chromatography methods for, 74
Equilibrium
 in chiral separations, 114–115, 117–118, 129–130, 132
 in drug detoxification, 27
Equilibrium constant, in chiral separations, 114, 130, 132
Esters
 of amino acids, 64–65, 68, 137
 chromatography methods for, 64–65, 71, 98–99, 139, 171
 high pressure liquid chromatography and, 200, 207
 as stationary phase, 188, 190–191
Etazepine acetate, preparative separations of, 155
Ethanol, as solvent, 76, 98, 139, 175, 187
Ethers
 chromatography methods for, 71, 98, 171
 macrocyclic (*see* Crown ether columns)
 as stationary phase, 104–105
Ethiazide, resolution of, 207
Ethosuximide, gas chromatography of, 71, 73
Ethylenes, polarized, molecular principles of, 45–46
Ethylurea, as stationary phase, 109
Etodolac, resolution of, 22–23, 205
Eudismic ratio
 adrenoreceptor comparisons, 17–20
 defined, 12
 laboratory variations of, 20–21
 Pfeiffer's rule for, 19
European Economic Community (EEC), regulatory guidelines of, 11, 30, 34
Eutomer
 defined, 12
 pharmaceutical applications of, 17–19
E,Z system, of configuration, 41

Fenfluramine, resolution of, 211
Fenoprofen, resolution of, 23, 205–206
Fischer, Emil, 46
Fischer projections, 46
Flame ionization (FID), for chiral detections, 82
Flow cells, for optical rotation detection, 88
Flow rate, effect on cyclodextrin separations, 182
Fluid chromatography
 subcritical, 76–77
 supercritical (see Supercritical fluid chromatography)
Fluorescence, for chiral detections, 59, 62, 83–84
Flurbiprofen, resolution of, 22–23, 175, 205–206
Food, Drug and Cosmetic Act, 34
Food and Drug Administration, U.S.
 pharmaceutical guidelines of, 30, 33–35
 regulatory summary of, 11–12
Formoterol, 19–21
Free energy difference, 117–120
Free solution capillary electrophoresis (FSCE), 77–78
Frontal chromatography, 146
Fruits, enantiomeric composition of, 212–213
Fungicides, preparative separations of, 153, 160

D-Galacturonic acid, as additive, 59
Gas chromatography (GC)
 chiral stationary phases for, 65–74: amino acid derivatives, 65–69; chiral metal complexes, 71–72; cyclodextrins, 68–71, 74–75, 208–209
 complexation, 71–74
 derivatization for, 64–65
 detection methods for, 82
 difficulties with, 63–64, 66
 mode selection in, 72–73
 natural compound applications of, 210–212
 overview of, 6–8, 63–64
 pharmaceutical applications of, 23–24, 211–213
 preparative separations for, 162, 164
 selected applications of, 208–212
G-BP (butyryl γ-cyclodextrin phase), 209
Glossary, of common terms, 12–13
Glutaramides, chromatography methods for, 66
Glycidyl 1-naphthyl ether (GNE), 26
Glycidyl 4-nitrophenyl ether (GNPE), 26
Glycoprotein columns, 123–125
G-PN (propionyl γ-cyclodextrin phase), 209

Gradient performance, from detectors, 84
Grains, enantiomeric composition of, 212–213
G-TA (trifluoroacetyl g-cyclodextrin phase), 208

Haüy, René-Just, 39
Helical chirality, preparative separations of, 153, 163
Helicenes, molecular principles of, 45–46
Helicity, stereoisomerism from, 41, 45, 68
Heptakis derivative, 70, 210, 213
Herbicides, preparative separations of, 153, 160
Herbs, enantiomeric composition of, 212–213
Herschel, John Frederick, 39
Heteroatoms, chiral, 153, 161
Hexakis derivative, 70
Hexane, as solvent, 27, 95, 104, 137–138
Hexobarbital
 comparative separations of, 216–218
 gas chromatography of, 71, 74, 211
 high pressure liquid chromatography of, 134–135, 203–205
High pressure liquid chromatography (HPLC)
 capillary electrophoresis versus, 78
 chiral mobile-phase additives in, 75, 132–135
 chiral stationary phase selection for: brush-type columns, 171–177; considerations for, 75–76, 170–171; cyclodextrin columns, 176–187; inclusion type, 176–187; ligand exchange type, 195–198; miscellaneous columns, 198–199; polysaccharide columns, 187–192; protein phases, 192–195
 detection methods for, 83–84
 of diastereomeric derivatives, 74
 mobile-phase systems in, 75, 100, 174–175, 178, 182–183, 186–187, 193, 198
 normal-phase, 135–136, 174, 182–184, 187
 overview of, 6–8, 74, 131
 reverse-phase (see Reverse-phased liquid chromatography)
 selected applications of, 170–212; analyte structure based, 199–207; stationary phases for, 171–199
 separation modes in, 131, 170–171
 solvents for, 148, 179–180, 187–189, 195, 221–223
 in supercritical fluid chromatography, 76
Hollow-fiber membrane reactors, 164, 167
Human serum albumin (HSA), as stationary phase, 194–195, 203

Hydantoins, chromatography applications of, 66, 106, 200, 205
Hydrocarbons, enantioselectivity of, 69, 71, 75, 121, 179
Hydrogen bonding, enantioselectivity of, 112–117, 121, 129–130
Hydrophobic interactions, in chiral separation, 115, 121, 123, 129–130, 138–139
Hydroxy acids
  derivatization of, 64–65, 102, 137, 171
  resolution of, 106, 132
Hydroxy compounds, chromatography and, 27, 199–201
Hydroxyl groups
  in chiral stationary phases, 96, 98
  chromatography methods for, 64–65, 74, 112, 129, 182
3-Hydroxy-3-phenylphthalimidine (HPP), 27
Hydroxypropyl β-cyclodextrin phase (B-PH), 208
Hypnotics, resolution of, 157, 200, 206–207

Ibuprofen, resolution of, 22–24, 175, 182, 194, 205–206
Ifosfamide, preparative separations of, 155, 157
Imides, high pressure liquid chromatography and, 200–201
Immobilization techniques, requirements of, 123
Imprinted polymers, as stationary phase, 105–106, 109, 130
Inclusion chiral stationary phase (Type 3)
  high pressure liquid chromatography applications, 176–187
  principles of, 97, 99, 120–122, 130, 134–140
  See also Cyclodextrins
Inclusion complexation, 178–179, 182
Indans, gas chromatography of, 71, 75
Indomethacin, 21
Inductive interactions, 129–130
Insecticides, preparative separations of, 153, 160
Intercalation, in chiral separation, 10, 114
Interconversion
  molecular principles of, 42, 45
  in pharmacokinetics, 27–28, 30
Inversion, in pharmacokinetics, 22, 27–28, 31
Ion complexes, transition metal, 42, 71–74, 102, 122–123
Ion exchange separations, 102, 105, 111, 122–123, 129–130
Ion pairing chromatography, 102, 105, 111, 122–123, 133–134, 151
α-Ionone, resolution of, 210
Isoborneol, resolution of, 210

Isoleucine, resolution of, 65
Isomenthone, resolution of, 210
Isomeric compounds
  biological activities of, 3–6, 53–55
  molecular structure of, 41–45
  separation flowchart for, 170–171
  stereoisomerism conventions of, 12, 41
Isomeric purity. See Enantiomeric purity
Isomerization, influencing factors of, 29
Isoproterenol, 54
Isozyme selectivity, in drug interactions, 26–29

Ketamine, 16, 154
Ketones, chromatography methods for, 65, 71, 171
Ketoprofen, resolution of, 23, 175, 206
Ketorolac, 23
Kinetic resolution, for chiral separations, 6

L- conventions, defined, 12, 46
l rotation. See Levorotation
Labetalol, 19–20
Lactones, resolution of, 106, 171, 211
Lactose, as stationary phase, 96
Leucine
  resolution of, 60, 62, 66–67
  as stationary phase, 109, 172
Leucovorin, resolution of, 205
Levomethorphan, 5, 14
Levorotation (l), defined, 4, 12, 85, 128
Lifibrol metabolite, preparative separations of, 157
Ligand exchange columns (LEC) (Type 4)
  for diastereomeric complex formation, 133
  high pressure liquid chromatography and, 195–198
  stationary phase principles of, 99, 102, 109, 136, 140, 150–151
Ligand exchange plates, for chromatography, 60, 63–64
Ligand formations
  in stereoisomeric interactions, 122, 125, 129–130, 136
  in transition metal complexes, 42, 71–74, 102, 122, 133
Ligand locations, nomenclature for, 41, 47
Limonene, resolution of, 210
Linalool, resolution of, 210
Lipodex stationary phases, 71
Liquid chromatography
  high pressure (see High pressure liquid chromatography)
  ion pairing, 102, 105, 111, 122–123, 133–134, 151
  low pressure, 103, 123
  reverse-phase (see Reverse-phase liquid chromatography)

Lorazepam, preparative separations of, 158
Low pressure liquid chromatography, 103, 123

Macrocyclic antibiotics
  cyclodextrin column separations of, 184–187
  stationary phase principles of, 121, 137, 215
Macrocyclic polyethers. *See* Crown ether columns
Magnetic resonance, nuclear, chromatography applications of, 115, 125
Maltodextrins, in capillary electrophoresis, 215–216
Malus, Étienne-Louis, 39
Mandelic acid, resolution of, 135
Marketing, of single-isomer pharmaceuticals, 15–17
Mass balance, in pharmaceuticals, 28–29
Mass spectrometer (MS), as detector, 83–84
MECC (micellar electrokinetic capillary chromatography), 78
Mefloquine (MQ), 25
MEKC (micellar electrokinetic chromatography), 77–78
Membrane separations, overview of, 6–8
Menthol, resolution of, 210
Mephenytoin, chromatography of, 134–135, 205, 211
Mephobarbital (MPB), chromatography of, 25, 71, 74, 203–204, 211
Mesitylene, as solvent, 122
Metabolism
  of drugs (*see* Pharmacokinetics)
  oxidative, 26
Metal coordination complexes, in gas chromatography, 42, 71–74, 122
Metal ion alkene coordination complexes, 42
Metal ion complexes, transition, in chromatography, 42, 71–74, 102, 122, 133
Metallocenes, molecular structure of, 42
Methanol, as solvent, 95, 100, 104, 139, 153, 175, 183, 187
Methaqualone, preparative separations of, 154
Methionine, resolution of, 60, 62
Method development
  chromatographic methods, 130–131
  stereoisomeric interactions, 128–130
Method selection
  for capillary electrophoresis, 214–216
  and comparative separations, 216–218
  for gas chromatography, 208–212
  for high pressure liquid chromatography, 170–207
  molecule scrutiny for, 169
  options for, 170
  starting point for, 168–169
  for supercritical fluid chromatography, 212–214
  tutorial for, 79–80, 144
Methylene chloride, as solvent, 175–176
Metoprolol, resolution of, 201–202
Mianserin, preparative separations of, 154
Micellar electrokinetic capillary chromatography (MECC), 78
Micellar electrokinetic chromatography (MEKC), 77–78
Microextraction, solid-phase, 212
Mirror images, of objects, 3, 38–40, 52
Mobile-phase systems in chiral separation
  additive method, 75, 132–135
  column selection procedure for, 137–140
  compatibility of, 95
  high pressure liquid chromatography and, 75, 100, 174–175, 178, 182–183, 186–187, 193, 198
  principles of, 7–9, 49–50, 58
  supercritical fluid chromatography and, 76–77
  thin-layer chromatography and, 59–64
  solvents for, 95, 104, 137–140, 148
Modifiers, effect on protein separations, 193–195
Molecular docking processes, 125, 140
Molecular modeling, chromatography and, 115, 125, 130
Molecules. *See* Chiral molecules; Template molecules
Monomers, defined, 10
Monoterpenes, preparative separations of, 162
M,P system, of configuration, 41

Naproxen, resolution of, 175, 206
Natural compounds
  gas chromatography of, 210–212
  as pharmaceuticals, 16, 212
Nefopam, preparative separations of, 154
Neutron diffraction analysis, for absolute configuration determination, 53
New chemical entities (NCE), guidelines for, 30–33
New drug applications (NDAs), guidelines for, 33–35
Nicardipine, 26
Nitrates, chromatography methods for, 98
Nitrogen compounds, basic, high pressure liquid chromatography and, 199, 201
Nomenclature systems
  D- or L-, 12, 46
  E,Z, 41
  M,P, 48
  R or S (*see* Cahn-Ingold-Prelog Convention)
  in stereoisomerism, 12, 41, 46–48

Nonsteroidal anti-inflammatory drugs (NSAIDS)
  enantiomers of, 21–22, 138
  pharmacokinetics of, 23–24
  resolution of, 205–206
  See also specific drug
Noradrenaline, 19
Norbormanols, resolution of, 67
Norepinephrine, 54
Normal-phase analyses, of chiral stationary phases, 135–136, 174, 182–184, 187
Nuclear magnetic resonance (NMR), chromatography applications of, 115, 125

Octanis derivative, 70
Olefins, chromatography methods for, 74
Omeperazole, preparative separations of, 157
One-point binding model, 130
Optical purity, determination of, 52, 87. *See also* Enantiomeric purity
Optical resolutions. *See* Chiral separations
Optical rotation
  in absolute configuration, 52–55
  chirality applications of, 49–52, 81, 85
  defined, 3, 12
  detectors for, 85–88
  discovery of, 39, 49
  enantiomer properties of, 3–5, 40
  negative molecular (*See* Levorotatory)
  positive molecular (*See* Dextrorotatory)
  See also Resolution
Optical rotatory dispersion (ORD)
  advantages of, 86–87, 90–91
  as chiroptical technique, 85–86, 88
  defined, 50, 52
  sample detectability, 87–88
  spectrum detection comparisons, 88–90
Overhauser effect, intermolecular nuclear, 115
Ovomucoid column (OVM), 123–124
Oxapadol, preparative separations of, 154
Oxazepam (OX)
  chromatography methods for, 27, 29, 76–77, 214
  preparative separations of, 156–157
Oxidative metabolism, 26
Oxindanac, preparative separations of, 154
Oxyprenolol, preparative separations of, 156

Paper chromatography. *See* Cellulose
Partition coefficient differences, in stereoisomeric interactions, 129
Partitioning, for chiral separations, 6
Pasteur, Louis, 40, 128
Peaks, in chiral stationary phases, 92–94, 209, 211

Penflutizide, preparative separations of, 155
D-Penicillamine, 16, 29
Pentane, in chromatography, 76
Perchloric acid, 105
Permethylated b-cyclodextrin phase (B-PM), 208
Pesticides, preparative separations of, 153, 160
Pfeiffer's rule, for eudismic ratio, 19
pH
  effect on cyclodextrin separations, 181–183, 187
  effect on protein separations, 193, 196–197
  racemization and, 29
Pharmaceutical Manufacturers Association, 30
Pharmaceuticals, chiral
  α1-acid glycoprotein column resolution of, 192–194
  capillary electrophoresis of, 215–218
  chromatography methods for, 59–60, 66
  configuration studies for, 52–55
  cyclodextrin separations of, 184, 186–187
  detection methods for, 81
  detoxification studies of, 26–30
  gas chromatography of, 23–24, 211–213
  industrial perspectives, 30–33
  interaction studies of, 26–29
  overview of, 4–6, 14–16
  pharmacokinetic studies of, 23–26, 31–32, 35
  preparative separations for, 153–159
  racemic, 5, 15–18, 26, 29, 31–33
  regulatory guidelines for, 27–28, 30–31, 33–35
  single-isomer: development of, 16–23; marketing status of, 15–17
  solvent factors, 27, 29
  stability studies, 27–30
  stereoisomers as, evaluation of, 14–37
  supercritical fluid chromatography of, 214
  synthetic versus natural, 16
  teratogens, 5, 14
  See also specific drug
Pharmacokinetics
  assay guidelines, 31–32, 35
  stereoselective, 23–26
  study methods and results, 23–26
Phases, in chiral separation
  biopolymeric (*see* Protein columns)
  cyclodextrin, 208–209
  dipeptide, for gas chromatography, 66
  inclusion, 97, 99, 130, 134–135, 176–187
  Lipodex stationary, 71
  mobile (*see* Mobile-phase systems)
  normal-phase analyses of, 135–136, 174, 182–184, 187
  rational stationary, 10

Phases (*continued*)
  reverse-phase analyses of (*see* Reverse-phase liquid chromatography)
  stationary (*see* Chiral stationary phases)
Phenobarbital, chromatography of, 25, 134–135
Phenylethanolamine, 18, 20
Phenylglycine, as stationary phase, 106–107, 109, 114
Pheromones
  chromatography methods for, 71
  preparative separations of, 153, 160, 162
Phosphates, in chromatography, 182
Phosphorus compounds
  high pressure liquid chromatography and, 200, 207
  preparative separations of, 153, 161
Phthalide, as stationary phase, 106
π acids, as stationary phase, 150–152
π bases, as stationary phase, 150–152
π-electron systems, in chromatography, 115, 121
π-π bonding, in chiral stationary phases, 106, 113–114, 129, 138
Pindolol, resolution of, 193, 202
α-Pinene, resolution of, 210
Pirkle columns, covalent, 27
Pirkle-type columns, 137–138, 204
Pirprofen, resolution of, 23
Planar chiral compounds, preparative separations of, 153, 162
Planar chirality, defined, 41, 46
Plane-polarized light, chirality applications of, 50–51, 128
Plasma concentration time curve (AUC), 23–25, 32
Plasma concentrations, of drugs, pharmacokinetic principles of, 23–26
Plates, for chiral separations
  commercial chromatography, 60, 63–64
  desirable features of, 93–94
Polar functional groups
  cyclodextrin columns and, 179–180, 182, 186
  derivatization of, 137
  polysaccharide columns and, 188, 191
  See also specific group
Polarimetry, 49–52, 85–88, 90
Polarized light
  chirality applications of, 49–52, 85–86
  circularly-, 85
  discovery of, 39–40, 49
  plane-, 50–51, 128
Polyacrylamides, as stationary phase, 102–104
Polymerization, suspension technique for, 102–103

Polymers
  imprinted, 105–106, 109, 130
  pharmaceutical applications of, 214
  polymethacrylate, 102–104, 199
  stability of, solvents and, 103–104
  as stationary phase, 102–106, 109, 150
Polymethacrylate columns
  high pressure liquid chromatography and, 199
  stationary phase principles of, 102–104
Polysaccharide columns (Type 2)
  categories of, 187–188
  high pressure liquid chromatography and, 187–192
  selection per functionality, 188, 191
  stationary phase principles of, 96–100, 109
Praziquantel, preparative separations of, 155
Precolumn derivatization, in chiral separation, 8, 74, 131–132
Prelog, Vladimir, 50
Preparative separations
  challenges with, 11, 79, 95, 102
  chiral stationary phases for, 148–151
  column sources for, 221
  desirable features of, 150
  displacement chromatography and, 146–147
  for drugs, 153–159
  elution chromatography and, 95–96, 147–167
  frontal chromatography and, 146
  gas chromatography and, 64
  large-scale systems of, 10–11, 79, 130–131, 163
  method development for, 152
  overview of, 10–11, 145–146
  performance of, 150
  resolution of, 148, 151–152
  selection of chromatography mode for, 145–146
  solvents for, 148–149, 152–153
  tutorial for, 167
Prisms, in polarimetry, 50–51
Product development guidelines, for pharmaceuticals, 30–35
Proglumide, preparative separations of, 158
Proline, resolution of, 67, 151, 172
Promethazine, 16
Propanolol
  electropherogram of, 215–216
  enantiomers of, 16–17, 132
  pharmacokinetics of, 24, 26
  preparative separations of, 156, 172–173
  resolution of, 194, 202, 215–216
Propeller chirality, preparative separations of, 153, 163

Propionic acid, as stationary phase, 106–107
Propionyl γ-cyclodextrin phase (G-PN), 209
Prostaglandin synthetase, 22
Prostaglandins, preparative separations of, 157
Protein binding
   chromatography applications of, 101
   pharmacokinetic principles of, 23–26
Protein columns (Type 5)
   chromatographic separations on, 122–125
   high pressure liquid chromatography and, 192–195
   silica-binding properties of, 123–125, 171–172, 194
   stationary phase principles of, 101, 109, 122–125, 136, 140, 152
Protein racemization, 29
Purity. See Enantiomeric purity
Pyrene, degree of fit of, 179

R conventions. See Cahn-Ingold-Prelog Convention
Racemate-enantiomer switches, 32–33
Racemates
   defined, 4, 128
   development of, 30, 32
   pharmaceuticals as, 5, 15–18, 28–29, 32; industrial perspectives of, 30–33; regulatory guidelines for, 33–35
Racemic compounds
   high pressure liquid chromatography and, 192, 200, 207
   pesticides as, 153, 160
   pharmaceuticals as, 5, 15–18, 26, 29, 31–33
   resolution into enantiomers, 145
Racemic solutes, resolution of, 106
Racemization
   with gas chromatography, 64–65, 210
   ibuprofen patterns, 23–24
   optical rotation detection of, 86–87
   predictive rules for configurational instability, 78–79
   of thalidomide, 14–15
Radiochemical detection techniques, 81
Rational stationary phase, study approaches for, 10
Reagents, in chiral separation, 8–9, 131, 137
Reciprocity, in chiral recognition models, 113, 115
Recycling, of solvents, 149, 159
Refractive index detection method, 83–84
Regulatory guidelines
   of European Economic Community, 11, 30, 34
   overview of U.S., 11–12

   of pharmaceuticals, 27–28, 30–31, 33–35
   for toxic compounds, 35
Repulsive interactions, in chiral separations, 114–115, 130
Research strategies, pharmaceutical, 30–33
Resolution
   baseline calculation of, 92–94
   charge-transfer based adsorption for, 106
   defined, 48, 90
   degree of pharmacological, specification of, 15, 20–22
   enantiomeric (see Enantiomeric resolution)
   methods for, 6, 48–52
   optical (see Chiral separations)
   of preparative separations, 148, 151–152
   relative power of, in chromatography, 94–95
Resolution coefficients, of selector/-selectand systems, 66–68
Resonance, in stereoisomeric interactions, 94–96, 99, 101, 105, 107, 129
Restek columns, 211–212
Retention factor, in detection methods, 81–83, 216
Reverse-phase liquid chromatography (RPLC)
   with brush-type columns, 174
   chiral stationary phases in, 99–100, 121, 130, 132, 136, 151
   with cyclodextrin columns, 178–182, 187
   with protein columns, 194
Rigidity
   of molecules, 42
   in stereoisomeric interactions, 129
   See also Conformation
Rolipram, preparative separations of, 154
Rosemary oil, analysis of, 211

S conventions. See Cahn-Ingold-Prelog Convention
Salbutamol, 20, 54
Scrape-and-elute detection method, 81–83
Sedatives, resolution of, 157, 200, 206–207
Selectors
   in chiral separation, 8–9, 106, 131
   synthetic, 106
Selector/-selectand systems, resolution coefficients of, 66–68
Selegiline, resolution of, 209
Sensitivity, of detection methods, 81–83
Separation factor, of resolution, 93–94, 100
Separation methods, chiral, 6–8, 10–11, 49–52, 58. See also specific method

Silica
  for chiral separations, 9–10, 59–61, 66, 70, 97, 106, 136–139, 150
  in enantioselective interactions, 113–114
  protein-binding properties of, 123–125, 171–172, 194
Silicones, cyanopropyl, in gas chromatography, 68, 70
Simulated moving-bed chromatography (SMB), 152, 163, 166
Single enantiomers
  development of, 16–23
  marketing status of, 15–16
  with one chiral center, 17–19
  purity of, biological consequences and, 20–22
  toxicity potential of, 16
  with two chiral centers, 19–20
Sizing, for chiral separations, 49
Solid-phase microextraction (SPME), of natural compounds, 212
Solubility, of samples in mobile phase, 148
Solutes
  charged (see Ion pairing chromatography)
  in chiral separation, 8–9, 82, 87, 97
  protein column separations for, 123–124
Solvents
  for chiral stationary phases, 98–99, 101, 104–105, 107
  effect on protein separations, 193–195
  for high pressure liquid chromatography, 148, 179–180, 187–189, 195, 221–223
  for low pressure liquid chromatography, 103
  for mobile-phase, 95, 104, 137–140, 148
  for pharmaceutical studies, 27, 29
  polymer stability and, 103–104
  for preparative separations, 148–149, 152–153
  recycling of, 149, 159
  stereoselectivity and, 115, 117, 121–122
  for thin-layer chromatography, 59
Spatial relationships, in stereochemistry, 3, 38–39, 45–46
Specific rotation, of liquids, 85–86
Spectroscopy, 85
Sprague-Dawley rats, 23–24
Stability
  in chiral stationary phases, 94–96, 99, 101, 105, 107
  chromatography and, 66, 68, 71, 116, 123
  configuration factors, 12, 27–29, 43, 45
  in diastereomeric complexes, 9, 133–135
  enantiomeric, 12, 27–31, 41–46
  in pharmaceuticals, 27–30
  polymer factors of, 103–104

Starches
  for chiral separations, 9–10
  degradation products of, 120
  as stationary phase, 96–100, 192
Static stereochemistry, 40
Stationary phases. See Chiral stationary phases
Stereochemistry
  absolute configuration determination in, 52–55
  configuration principles, 41–46, 52–55
  conformation principles, 41–46
  defined, 38
  dynamic, 40
  history of, 39–40
  molecular classifications, 40–41, 45–46, 111
  nomenclature systems, 12, 41, 46–48
  pharmaceutical applications of, 14–35
  resolution principles of, 48–52
  static, 40
Stereogenic center
  defined, 3, 12, 40
  determination of, 169–170
  pharmaceutical consideration of, 27–28
  as stereoisomerism source, 40–43
Stereoisomeric enrichment, in pharmaceuticals, 28–29
Stereoisomeric interactions
  analytes in, 112–114, 120, 123
  attractive, 114–115, 138–139
  charge-transfer, 115–116
  for chiral separation, 114–115, 121, 128–129
  chromatography and, 111–114, 129–130
  developmental overview of, 128–130
  dipole-dipole, 114–115, 121, 129–130
  electrostatic forces in, 114, 129–130
  hydrogen bonding, 112–113, 115–117, 121, 129–130
  hydrophobic, 115, 121, 123, 129–130, 138–139
  inductive, 129–130
  interference in, 129
  ionic, 102, 105, 111, 122–123, 129–130
  ligand formations in, 122, 125, 129–130, 136
  partition coefficient differences in, 129
  π-π systems, 106, 113–114, 116, 121, 129
  repulsive, 114–115
  resonance, 94–96, 99, 101, 105, 107, 129
  structural rigidity in, 129
  temperature factor, 82, 86, 117–120, 129
Stereoisomerism
  pharmaceutical applications of, 23–29
  sources of, 3, 40–43

Stereoisomers
 defined, 3, 13
 as drug candidates (*see* Pharmaceuticals)
 optical rotation detection of, 87
 symmetry factor of, 3
Stereoselective synthesis
 chromatographic theories for, 111–120
 of enantiomers, 10–11, 49, 145
 pharmacokinetic applications of, 23–27, 29, 145
 resolution methods for, 86–87, 90
Stereoselectivity, solvent factors, 115, 117, 121–122
Steric crowding, in molecules, 45
Steric fit, in enantiomeric separations, 115, 130
Steric interference, in stereoisomeric interactions, 129
Subcritical fluid chromatography (SubFC), of oxazepam, 76–77
Substituents, molecular applications of, 42–45, 47
Subunits, of chiral separations, 10
Sulfoxides, resolution of, 106, 132, 153
Sulfur compounds, high pressure liquid chromatography and, 200, 207
Supercritical fluid chromatography (SFC)
 advantages of, 212
 alcohol effects on, 212–214
 overview of, 6–8
 pharmaceutical applications of, 214
 preparative separations for, 152, 163, 165
 selected applications of, 212–214
 separation process in, 76–77
Superimposibility
 of chirality, 3, 38–39
 molecular principles of, 42, 44
Suspension polymerization, 102–103
Symmetry factor, in stereoisomerism, 3, 40–41
Sympathomimetic amines, absolute configuration of, 54–55

Tailing, in chiral stationary phases, 93–94, 211
Tartaric acid, as additive, 60–61
Tartrate anion, configuration of, 53
Temperature
 as chiral separation factor, 63, 68, 117–120, 129
 effect on cyclodextrin separations, 182
 effect on ligand exchanges, 196, 198
 as gas chromatography factor, 63–64, 68–73
 in optical rotation detection, 82, 86, 117
 stereoisomeric interactions and, 82, 86, 117–120, 129

Template molecules, for polymer synthesis, 105–106
Teratogens, enantiomers as, 5, 14
Terbutaline, chromatography methods for, 17, 78
Terminology, glossary of, 12–13
Tetralins, gas chromatography of, 71, 75
Thalidomide
 preparative separations of, 156
 racemization half-life of, 14–15
 teratogenic activity of, 5, 14, 34
 twisted conformation of, 45
Theoretical plates, in chiral separations, 93–94
Thermal conductivity (TC), for chiral detections, 82
Thermodynamics. *See* Temperature
Thin-layer chromatography (TLC)
 achiral stationary phase/achiral mobile phase approach, 59–60
 achiral stationary phase/chiral stationary phase additives approach, 59–62
 additives for, 59–60
 chiral stationary phase/achiral mobile phase approach, 59–64
 detection methods for, 81–82
 overview of, 6–8, 59
 plates for, 60–64
 solvents for, 59
Thioethers, chromatography methods for, 71
Thiols
 chromatography methods for, 74
 resolution of, 106, 132
Three-point rule, of stereoisomeric interactions
 principles of, 96, 112, 126, 129
 resolution capacity per, 173–174
Tissue binding, of drugs, pharmacokinetic principles of, 23–26, 29
Torsional chirality, defined, 41
Toxic compounds
 enantiomers as, 4–6
 preclinical study guidelines for, 30, 32–33
 regulatory guidelines for, 35
 single-isomer drugs as, 16
Trace components, determination of, 95
trans ligand, 41, 47
Transition metal ion complexes, in chromatography, 42, 71–74, 102, 122, 133
Triacetate derivative of cellulose, for chiral stationary phase, 98, 151–153
Tribenzoate derivative of cellulose, for chiral stationary phase, 98
Tribenzyl ether derivative of cellulose, for chiral stationary phase, 98

Trichoderma reesci, 123
Tricinnamate derivative of cellulose, for chiral stationary phase, 98
Trifluoroacetyl γ-cyclodextrin phase (G-TA), 208
ρ-Trifluoromethyl anilide derivative (PTFMA), 19–20
Trimethoquinol, resolution of, 215, 217
Triphenylmethyl methacrylate, synthesis of, 103–104
Trisphenylcarbamate, for chiral stationary phase, 98
Troeger's base, resolution of, 201
Tropane alkaloids, 30
L-Tryptophan, resolution of, 123, 134, 210
Tutorials
　on chiral separations, 110, 126–127
　for chiral stationary phase selection, 144
　on chiroptical detectors, 90–91
　for chromatography method selection, 79–80
　for detection methods, 90–91
　for preparative separations, 167
　for understanding chromatography, 126–127
Twisted conformation, 45–46
2^n Rule, 169
Tyrosine, resolution of, 62, 76

Ultraviolet light (UV)
　for chiral detections, 59, 77, 81, 83–84
　detection spectrum comparisons, 87, 89–91, 199
Urea-type columns
　high pressure liquid chromatography and, 171–174
　stationary phase principles of, 109, 114

Valine
　resolution of, 66–68
　as stationary phase, 109, 151, 172–173
Vancomycin, as stationary phase, 184, 186–187
Volatility, gas chromatography and, 64–66, 71

Warfarin, resolution of, 26, 157, 193
Wavelengths, in optical rotation detection, 85–86, 88–90
Whelk-O 1 columns
　applications of, 200–201, 207
　principles of, 174–176

X-ray crystallography
　for absolute configuration determination, 53
　of diastereomeric complexes, 115–117
X-ray scattering, anomalous, 53